Construction Claims & Responses

effective writing & presentation

This book is dedicated to Kim, Nik and Nina.

Construction Claims & Responses

effective writing & presentation

Second Edition

Andy Hewitt

Hewitt Construction Consultancy Claims Class

WILEY Blackwell

This edition first published 2016
© 2016 by John Wiley & Sons, Ltd
© 2011 by John Wiley & Sons, Ltd

Wiley-Blackwell is an imprint of John Wiley & Sons, formed by the merger of Wiley's global Scientific, Technical and Medical business with Blackwell Publishing.

First edition published 2011
Second edition published 2016

Registered Office
John Wiley & Sons, Ltd, The Atrium, Southern Gate, Chichester, West Sussex, PO19 8SQ, United Kingdom

Editorial Offices
9600 Garsington Road, Oxford, OX4 2DQ, United Kingdom
The Atrium, Southern Gate, Chichester, West Sussex, PO19 8SQ, United Kingdom

For details of our global editorial offices, for customer services and for information about how to apply for permission to reuse the copyright material in this book please see our website at www.wiley.com/wiley-blackwell.

The right of the author to be identified as the author of this work has been asserted in accordance with the UK Copyright, Designs and Patents Act 1988.

Library of Congress Cataloging-in-Publication Data

ISBN: 9781119151852

A catalogue record for this book is available from the British Library.

Wiley also publishes its books in a variety of electronic formats. Some content that appears in print may not be available in electronic books.

Cover image: SilviaJansen/iStockphoto

Set in 10.5/13pt Helvetica by SPi Global, Pondicherry, India

1   2016

Contents

About the Author

Andy Hewitt is a construction contracts and claims consultant who has over 40 years' experience in the construction industry in the UK, Africa and in the Middle East. In addition to many projects in the UK, he has worked on projects in Nigeria, Sudan, Bahrain, Saudi Arabia, Jordan, Oman, Tanzania and the United Arab Emirates.

Andy has held senior commercial and project-management positions with contractors, subcontractors and consultants, including several years operating his own consultancy practice in the UK, which provided quantity surveying, estimating and project-management services. He has been involved in a wide variety of construction projects including super high-rise, airports, hospitals, residential, hotels, shopping malls, industrial buildings, heavy civil engineering, marine works, process plants, pipelines, desalination plants and royal palaces.

One of the most enjoyable periods of Andy's career was when he was employed as a claims and contracts consultant by J.R. Knowles, one of the most prestigious international consultants in this field. During this period he discovered that his background on both the contractor's and the client's 'side of the fence' in commercial, contracts and project-management positions gave him the ability to look at the issues objectively and to manage and to resolve the often adversarial nature of claims in a proactive manner to achieve resolutions acceptable to the parties. During this period he was involved in several iconic projects in the United Arab Emirates including the world-famous Burj Al Arab hotel in Dubai.

During the past decade, Andy has been employed in positions as project director, contracts manager and commercial manager on many prestigious projects in the Middle East. In these roles he has had the responsibility of both preparing and determining many claims. The often poor quality of the claims and determinations that have come across his desk during this time, many of which have exhibited a lack of understanding of many basic concepts and requirements of the subject, inspired him to share his experience on the subject with the industry by way of this publication.

Currently, Andy is the principal of Hewitt Consultancy services, a small, specialist practice offering claims, contracts, dispute management and adjudication services to contractors and employers.

Following feedback from the first edition of this book and after several invitations from companies to provide training on claims related issues, Andy realised that there was a significant gap in the market for education and training on the subject. This led him to develop a tutor-assisted distance learning course on construction claims and several two-day intensive training courses on subjects relating to claims. These courses are marketed internationally under the banner of Claim Class.

Andy now splits his time between Europe and the Middle East and is involved in consultancy work as well as providing education and training in many countries.

Hewitt Construction Consultancy provides two distinct services to clients: consultancy work, which is principally provided by Andy Hewitt, and education and training which is provided under the Claims Class banner.

Consultancy services include the following:

- Claim preparation
- Claim review and defence
- Claim enhancement and presentation
- Contractual advice
- Contract administration set up and advice
- Dispute adjudication
- Dispute management
- Arbitration support
- Dispute board membership
- Education and training

Further Information

Email: hello@hewittconsultancy.com
andy.hewitt@hewittconsultancy.com
www: www.hewittconstructionconsultancy.com

Foreword

I was pleased to be asked by Andy Hewitt, a former colleague, to write the Foreword to his book.

The first thing that strikes one, having read the first few pages, is the easy style Andy has adopted, which made my task a pleasure. It is also obvious from the outset that the author has had a great deal of hands-on experience of preparing and responding to claims, and this oozes from the pages.

First and foremost this book is international in its outlook and will be useful for those involved in claims on a worldwide basis. In the early part of the book, Andy recounts his need when preparing his first claim for a 'Claims for Dummies' type of book which he couldn't find. This is not a book for dummies, but is essential reading for anyone who is preparing a claim for the first time. For those of us with experience aplenty, the book provides an excellent aide memoire and will ensure that nothing is missed.

The book is without a doubt fully comprehensive and goes through the preparation of a claim from A to Z. In each chapter Andy tells the reader of matters which he intends to cover, then provides the detail and ends up reviewing what had been written. In any campaign, – and the preparation submission and negotiation of a claim is something of a campaign – it is essential to have a strategy, and this is dealt with at the outset.

Claims may relate to variations, delays caused by the employer and neutral events which could involve extensions of time, prolongation costs, acceleration and disruption, all of which are fully explained.

Claims are nearly always prepared by reference to the conditions of contract. The book refers to the FIDIC conditions, but this should not put off those who are involved with contracts where other standard conditions apply. The comprehensive nature of the book would easily enable the reader to slot the advice provided on its pages into other standard conditions of contract. The book leaves nothing to chance when referring to the conditions that are applicable when preparing a claim.

The need for a stand-alone claim, accompanied by all documents referred to therein, is stressed as being essential if the claim is to be taken seriously and to result in a satisfactory settlement. Nobody who has the task of reviewing a claim has the appetite for wading through mountains of files to find documents that relate to the claim. The claim must be user friendly and be in more than one volume to ensure that when reading the claim it is easy to follow documents to which the claim relates. These may seem fairly basic matters, but I would say that in excess of half the claims prepared fail to follow this simple procedure.

Andy goes on to deal with what he considers to be the essentials of a successful claim CEES Cause, Effect, Entitlement and Substantiation. By way of illustration the book provides in detail the CEES of a delay and disruption claim on an 84-dwelling project where six of the houses are delayed and disrupted by work undertaken on behalf of the employer on the access road. There is also an example claim of an extension of time and additional payment for prolongation arising from a variation in respect of the redesign of an electrical transformer room on a multi-storey project.

The style and formatting of the claim document is dealt with down to such detail as the content and layout of the cover to the claim. Finally, Andy deals, from his experience, with how a response to a claim should be undertaken in a professional manner.

I like the book and have no hesitation in recommending it to students, beginners, those involved on a day-to-day basis with time and cost on projects, as well as the seasoned claims consultants.

It will certainly have a place on my bookshelf to allow me, having prepared a claim, to ensure that I haven't missed anything.

Roger Knowles

INSTITUTE OF CONSTRUCTION CLAIMS PRACTITIONERS

Overview

People who deal with claims within the construction industry inevitably do so after qualifying in some other profession, usually engineering, design, commercial management, contracts management or project management. Launched in 2015, the Institute of Construction Claims Practitioners (ICCP) recognises that, in order to properly prepare, respond to or manage claims, a level of professional expertise must have been achieved within what has become a specialist sector of the industry.

Claims and subsequent disputes can run into huge sums of money and it is considered that the owners of such sums should be afforded a level of confidence that those responsible for dealing with such matters on their behalf, whether employees or consultants, are suitably qualified and experienced.

The ICCP's mission is to maintain a professional institute, whereby suitably qualified and experienced professionals are awarded a recognised qualification related to the claims discipline.

Professional Standards

The institute sets professional standards for its members to ensure that institute members are suitably qualified and experienced at the specific membership level awarded.

The institute also encourages and assists members to constantly improve their professional standards and knowledge and to strive to achieve membership at the next level within the institute.

Information Sharing

The claims profession, especially when compared to other professions within the construction industry, is in its infancy. Consequently, research and reference material is often in short supply. The institute maintains a knowledge centre of information where papers, case studies, articles and presentations are made available to the members for reference. Members also have access to information on relevant books and higher education and training courses. A members' magazine is also published which contains information relevant to this sector of the industry, and future plans include the launch of a members' discussion forum and an annual conference.

Criteria for Membership

There are three grades of membership within the institute:

Associate (AICCP)
Member (MICCP)
Fellow (FICCP)

The level of membership is dependent on qualifications in other industry disciplines, together with specific and verifiable experience within the claims sector of the industry.

Further Information

Institute of Construction Claims Practitioners
Email: hello@instituteccp.com
www: http://instituteccp.com

CLAIMS CLASS

After the first edition of *Construction Claims & Reponses* was published, I was pleasantly surprised at how well it was received and by the fact that many readers took the trouble to contact me to say how useful that they had found the book. In many instances, people also inquired if I could recommend any educational or training courses to further enhance their knowledge of the subject. At the time, I was obliged to advise that, while there were several university degrees and training courses on construction or contract law, as far as I was aware, there was nothing that provided education or training specifically on the practical aspects of claim and response preparation as is covered in this book. The realisation that there was a significant gap in the market and that there was a need for education and training on the subject inspired me to take the subject of the book to the next level and consequently, Claims Class was born.

Claims Class now offers various education and training courses on matters associated with construction claims including:

Construction Claims Distance Learning Course

This tutor-assisted course comprises seven modules that cover all the subjects included in this book. The course requires an average of 150 hours of study and students can take as long as they wish to complete the course. Tutorials are provided to assist the students to complete each module and each student is assigned a tutor who acts as a mentor, grades assignments and provides feedback and advice. After the completion of each module, students are provided with model examples of how the assignment work could have been completed. The course examples provide a valuable source of future reference material.

Claims Class has so far had students enrol on the course from more than twenty-five nationalities from all over Europe, North and Central America, Eastern Europe, the Middle East, Africa, South East Asia, China and Australia. Feedback from those who have completed the course has always been excellent.

Intensive Training Courses

Claims Class offers two-day courses on subjects that are aligned with construction claims, including claims and responses, delay analysis and the FIDIC contracts. The exact subject matter is changed in accordance with demand. These courses are presented by highly qualified and experienced presenters and are broken down into various modules. In order to enhance the learning experience, delegates are divided into groups that are required to study and discuss various assignments throughout the course and present the findings to the class for further discussion.

The courses take place in various international locations, and Claims Class has presented such courses to hundreds of delegates in locations such as Abu Dhabi, Bahrain, Doha, Dubai, Dublin, Johannesburg, London and Singapore. Claims Class is always looking for opportunities to present these courses in new locations. As with the distance-learning course, delegate feedback is always excellent.

In-House Training

When companies have a requirement for training, it is sometimes more convenient and economical for Claims Class to go to them. Claims Class can either offer an existing training course to be delivered at the client's premises, or tailor-make a bespoke course to cover the specific requirements of the client.

Further Information

Claims Class
Hewitt Construction Consultancy
Contact: hello@constructionclaimsclass.com
Website: constructionclaimsclass.com

Acknowledgements

I would like to acknowledge the following people's help, encouragement and assistance:

Roger Knowles for encouraging me to undertake this publication, for his introduction to Wiley-Blackwell and for kindly writing the foreword.

The Knowles Group for providing me with some of the most rewarding and enjoyable years of my career and for the opportunity to learn the craft of claims.

I also wish to thank the Fédération Internationale des Ingénieurs-Conseils (FIDIC) for kindly allowing me to reproduce sections from their publication *Conditions of Contract for Construction for Building and Engineering Works Designed by the Employer, First Edition 1999.*

In this book, the Employer, the Engineer, the Contractor and Subcontractors are referred to in the masculine gender in conformity with standard FIDIC practice. The author wishes to emphasise that the book is intended to address female readers on an equal basis with their male colleagues and that all references throughout the book to the masculine gender are only for convenience in writing.

Chapter 1

Introduction

Why is it Necessary to Produce a Fully Detailed and Professionally Presented Claim or Response?

Imagine that you have been invited to an interview for a new job. This job is a real step up the career ladder and could enable you to move to a better house in a new area where your children would be able to attend a really good school. The job would be stimulating and interesting, and this is the chance that you have been waiting for to prove yourself professionally. When the time comes for the interview, you would undoubtedly take care of your appearance – wear a good suit and ensure that your shoes were polished. You would also probably have spent time thinking of how best to convince the interviewers that you are the ideal person for the job and would have rehearsed answers to the questions that you expect to be asked. If you thought that there may be some negative aspects to your qualifications or experience, lack of specialised experience for part of the job for example, you would probably have thought about how you could put a positive perspective to the interviewers, maybe by stressing some other aspect of your experience which could be easily drawn upon to overcome the perceived disadvantage. In short, given the rewards for success arising from the results of the interview, any person in this position would do their very best to sell themselves to the people making the decision.

Why then, do many of those people or companies responsible for presenting or reviewing claims, which often equate to considerable sums of money, not take similar pains to ensure that their submissions are presented in a professional and thorough manner; that they contain all the relevant and necessary information; and that they answer questions that will probably be asked by the reviewer of the claim? During the past several years, I have spent a considerable proportion of my time reviewing claims, and I can honestly say that during this time I have received very few submissions for which I have not had to raise queries or request additional particulars to be submitted. In some cases

Construction Claims & Responses: Effective Writing & Presentation, Second Edition. Andy Hewitt.
© 2016 John Wiley & Sons, Ltd. Published 2016 by John Wiley & Sons, Ltd.

I have simply rejected the claims as presented because they do not fulfil the basic requirements to prove that, on the balance of probabilities, the claim has any merit. Many of the submissions have consisted of a two-page or three-page letter enclosing a haphazard, dog-eared collection of documents which leave the reviewer to try to follow the logic of the claim and make his own conclusions. Well, I am sorry, but my job in such circumstances is to produce a determination, and however impartial I try to be, it is human nature that if my life is made difficult or if I am expected to do the claimant's work for them, I am hardly likely to be predisposed to giving the benefit of the doubt to the person who has brought about this state of mind. I probably would not offer a job to someone who turned up for an interview in a pair of scruffy old jeans with no real idea of how they could make a success of the position either.

Taking the point of view from the other side of the fence (and I am blowing my own trumpet a little here), I have also put together sound claim documents with reasoned and substantiated arguments to demonstrate clear entitlement and quantum, only to have them rejected out of hand by way of a few sentences with no real reasons being given for the rejection. Such responses are, if anything, even less professional than producing a bad claim document, because they are likely to lead to a costly dispute.

The aim of a claim is to persuade the respondent that, on the balance of probabilities, the claimant has entitlement under the contract and/or at law, and to succeed in this, the facts of the events on which the claim is based need to be presented in a logical manner and they need to be substantiated. The contract and, if necessary, the law should be examined to demonstrate that the events give cause to entitlement. If the claimant has been at fault in any way or if there are weaknesses in the claimant's case, then these should be considered and arguments made as to why such things should not affect the case. The same considerations should be taken into account when reviewing a claim. Has the claimant reasonably proved each element of his case and is he entitled to an award of all, or possibly part, of the claim and if not, why not? Such determinations should be clearly written in the response and the reasons for the conclusions should be adequately demonstrated and substantiated. The respondent needs to be equally persuasive that his arguments are well founded if he is to convince not only the claimant, but also the project owner of his findings, which he will need to do if the situation is to be settled and a dispute avoided.

Salespeople are taught that the first rule in selling is to get the customer to like them – the phrase used is 'selling yourself' and the principle here is that people do not want to buy from people that they don't like. We have all been in a situation whereby we grudgingly buy something from an obnoxious salesperson because we have no other choice and also in situations where we walk away as satisfied customers when we have been treated well by a friendly, helpful and likeable salesperson who is knowledgeable about their product. Presenting a claim or a response to a claim is exactly the same because the goal is to 'sell' it to the other party. The major difference, however, is that we have to initially promote our arguments by means of the printed word, so anything we can do to help 'sell' the claim by obtaining sympathy from the reviewer and by making it easier to agree has direct benefits on the likely outcome.

The Purpose of the Book

This book presents a guide to preparing claims in order to ensure that a claim submission contains all the relevant and necessary information to prove the case and that the document is set out in an easily understood manner, which leads the reviewer to a logical conclusion to demonstrate the claimant's entitlement. It also deals with the preparation and presentation of responses and determinations in a similar manner in order to resolve matters rather than prolong them. In addition to ensuring that the claim or response document contains all the necessary information, the following chapters also contain guidelines for making such documents user-friendly so as not to alienate the reviewer and thus make him less predisposed to disagree with your point of view.

In my experience, many problems surrounding claim and response writing arise from the fact that, in many cases, the people tasked with writing the claim or the response have little or no specific experience in the subject and tend to have to make things up as they go along. On a typical project the number of claims presented may be small and infrequent, so the persons responsible for producing the claim or response do not get the opportunity to practise the art on a regular basis. Adding the considerable requirements in terms of time and effort necessary to prepare a successful claim or a professional response to the pressure of their day-to-day tasks, there is little wonder why the result is often less than perfect.

A mechanical engineer is able to calculate the cooling requirements of an air-conditioning system through a proven and established set of rules to take into account the local climatic conditions, the thermal characteristics of the structure and the building's usage. Quantity surveyors use established and prescribed ways of calculating quantities so that any other person with the required level of experience can understand the methodology and verify the calculations. These tasks are possible because established procedures are taught either at learning institutions or on the job and are often recorded in industry publications. I believe that the writing of claims can be approached in the same way and this book attempts to set out a framework – a workshop manual, if you like – to assist claim writers and claims' reviewers in these tasks.

We usually think of claims in terms of a submission by a main contractor to the party responsible for administering the contract or by a subcontractor to a main contractor. While this is inevitably the first step, it is possible, however, that the situation may develop into a long and drawn-out affair involving many people and parties with varying degrees of skill and experience in such matters. While this book is primarily aimed at the project personnel responsible for writing or responding to claims, it is also hoped that the principles discussed and the examples worked through will be of use to those who deal with such matters on a more regular basis – dispute-review board members, mediators, adjudicators, arbitrators and the like.

In subsequent chapters, we will discuss the various types of claim and how they should be presented. We will consider what information should be included in claim documents and why. We will examine how to set out and present a claim in various sections so as to bring clarity to the presentation. Finally, we will discuss how to present the actual document in terms of a list of contents, page layout, appendices, exhibits and the like within the submission.

Chapter 2 contains some advice on contract administration for claims and claim avoidance. Chapters 3–5 are included to discuss claims in general terms and the key elements that are required to produce a successful claim submission together with examples of how various claims could be presented. We then move on to Chapters 6–9 in which we will discuss the claim document itself in detail and examine it section by section to discuss the purpose of each section, its content and the conclusions we are trying to reach. In these chapters, we will use a fictitious project in which an event has occurred which gives rise to a claim for an extension of time and additional costs and we will gradually build up a full example claim submission in order to provide examples of the wording, language and content.

Every claim will at some time require a response or determination and Chapter 10 provides guidance and an example of a response to a claim, in order to demonstrate how a professional and comprehensive response document should be produced. Finally, Chapter 11 provides some background and advice on dispute boards.

Although some basic legal precedents are relied upon in the example claim, it is not the purpose of this book to attempt to examine case studies, legal precedents or the like. In my experience the average claim at project level can get along quite nicely without resorting to complicated legal arguments and such matters generally only need to be brought into play in order to reinforce an area of the claim where there may otherwise be doubts as to entitlement, or if the claim evolves into a dispute. Having said that, there are some basic legal principles that often give strength to certain assertions and it is an advantage to have knowledge of such. It is therefore a good idea to have at hand legal references that have been applied to the construction industry and there are many excellent publications available for this purpose.

Things to be Considered Before Writing the Claim

When an event or events have occurred that the project team consider give rise to the need to submit a claim – typically an act of prevention by the Employer or his agents, or an event outside the control of the Contractor – it is sensible to consider certain matters before proceeding. Things to be considered in the early stages are the following:

1. The likely outcome and seriousness of the event. Will it have a serious enough impact on the claimant to justify the submission of a claim?
2. The value of the claim. Obviously the ends must justify the means here and if it will cost a significant amount to prepare a claim for a small return, it may make little economic sense to pursue the action. Similarly, if a claim is likely to attract a high return, it is probably worth providing the necessary resources to ensure a high-quality effort is made.
3. The strength of the claim and its chances of success. Are the odds of success great enough to justify the effort and expense?
4. The strategy should also consider how the claim is to be pitched. In general terms it is extremely unlikely that the claimant will receive the full value of his claim submission, so it is wise to include a negotiation margin. Is it therefore considered that the best result would be obtained by maximising all issues leaving a large amount for negotiation, or would it be better to ensure that all arguments are absolutely sound and the case is as bulletproof as possible? The latter usually results in a submission with a lesser value, but is often harder to defend against or to refute. It is also true to say that while an inflated claim strategy may have a chance of success if reviewed by inexperienced parties, if the matter subsequently proceeds to a dispute, such a claim is unlikely to succeed when the experts get involved.

5. Some claims are complicated in their very nature and if this is the case, they require a certain amount of knowledge and experience to prepare. Do in-house resources contain adequate experience and knowledge to produce the desired result, or should additional resources be brought in?

6. Client relationships should be considered. Claims are inevitably viewed negatively by the respondent. At best, the person responsible for reviewing the claim and making a determination will consider another task to include in his already busy working schedule as an inconvenience and at worst, the claim will be viewed as an attempt by the 'greedy and unscrupulous' contractor to maximise his returns by any means from the 'poor, hard-done-by and totally innocent' client. While the reality will most likely fall somewhere between these extremes, existing and future client relationships should be considered, maybe at executive level, before embarking on a course of action that could possibly end in contention.

7. The parties who are likely to make the determination should also be considered. Will they be difficult to persuade? Do they have a responsibility to protect the Employer's interests or to be impartial? If it is the latter, are they actually likely to act impartially? The actual personnel should also be considered. Has animosity crept into the relationship? Is the person likely to have sufficient knowledge to understand the matter in question and the contractual principles relied upon? Is the Employer likely to engage the services of an expert?

It is a good idea to develop a claim strategy in the very early stages of a project. The strategy should take into account the above considerations and decide upon what the claimant really wishes to achieve from the situation.

A good illustration of such considerations is that on a particular occasion I was engaged as a consultant to prepare an extension-of-time claim for the Contractor. The project was almost at an end when I was consulted and I was asked to compile a claim based on approximately twelve events that had delayed the Contractor. The claim was duly prepared and submitted to the Resident Engineer who was the party responsible for issuing extensions of time. After about a month, the Contractor and I were asked to attend a meeting with the Resident Engineer to discuss the matter. The person responsible went through each delay event in turn and gave his comments and I was pleasantly surprised that he was in agreement with most of the claim as submitted. I was even more surprised when, in a couple of instances, he even pointed out that we could have actually claimed more time. In two instances, however, he put up a vigorous argument as to why the Contractor was not entitled to anything at all for the events claimed.

The penny eventually dropped when I realised that all the events with the exception of the two that were being disputed could be laid fairly and squarely at the door of the Employer or on other external circumstances and that the two that were being vigorously defended were the fault of the Resident Engineer. The message was simple. Go away, revise your claim to make sure that the Resident Engineer appears blameless, resubmit it and I will then issue your extension of time. Professional? Not really, but we got the result we were looking for and that is what counts.

The contract administration procedures prior to submitting a claim should take into account that early communication is an important factor in influencing how a claim will be received. If the receiving party is taken by surprise when the claim lands on his desk, he is likely to feel 'ambushed' or even consider himself as being professionally inadequate because he failed to see it coming and probably did not report it to his client or superiors. It is consequently quite natural for a person in such a situation to offer up a rigorous defence. On the other hand, if the party has been forewarned through formal and informal communications that the claimant considers that he has an entitlement to a claim and that a submission will be forthcoming, then the recipient will not only be mentally prepared for its arrival, but he should have made adequate provisions for it in his reports, budgets and the like. It is more likely that a reviewer will adopt a more impartial position if financial provisions have already been made against a possible claim, than if he has to go cap in hand, bearing bad news to those further up the tree who could possibly adopt a 'shoot the messenger' mentality in such a situation.

When considering the matter of the person or persons responsible for preparing the claim, it is definitely worth thinking about consulting with an expert if one does not exist on the project. In an earlier life I worked as the project manager for a subcontractor who was delayed significantly on the project by the main contractor and this brought about my first experience at claim writing. I had no one to help or to advise me and, try as I might, I could not find a 'Claims Writing for Dummies' type of book to give me any guidance, which of course is something that I am attempting to rectify by this publication. I struggled on through and ended up with what I considered at the time to be a decent submission. Looking back nowadays, however, I realise that it was not a very professional effort and I really could have benefited from some advice from someone who had been there, done it and got the T-shirt. Similarly, having worked in various positions on projects on which I have had the responsibility of dealing with claims and where there are never enough hours in the day, I know from personal experience that day-to-day life on site leaves little time to sit down and put in the necessary concentrated effort to produce a robust claim submission or response. In a later part of my somewhat varied career, I had the good fortune to work for the Knowles Group as a consultant, and in those days, the main purpose of my life was to either prepare claims on behalf of contractors and subcontractors, or to review claims and advise on determinations on behalf of employers. I have to say that because I was doing this type of work on a day-in, day-out basis, I developed efficient ways of doing things due entirely to constant practice and the consequent economies of scale. I would not say that it became a production line exactly, but repetition certainly improved my ability to produce the work efficiently and effectively. Having come to this job from various project-management or commercial-management roles, I also found that it was a luxury to be able to sit down without the phone ringing every five minutes or people constantly popping into my office needing something attending to yesterday. In short, in those days I was able to concentrate fully on the task in hand, which is often not something that most project personnel are able to do.

One other advantage of bringing in an 'outsider', whether from company resources or through the services of a consultant, is that because the project personnel live and breathe the project on a day-to-day basis, it becomes very personal to them. Personalities consequently tend to come into play and emotions tend to surface, and it is sometimes therefore difficult to view a situation in a detached and objective manner. In my experience the 'outsider' will often be able to take a few steps back when examining a situation and he will consequently be able to give a more dispassionate opinion on the merits of the claim and possibly cut right to the bare bones of the matter to produce a more balanced and less emotive submission than might otherwise have been the case.

The form of Contract used in the Examples

As we progress through the book and discuss the process of building up and writing claims, it will be necessary to refer to various conditions of contract. There are many standard forms of contract in use internationally and it is not the purpose of this book to examine such contracts, but rather to explain how to use a *typical* contract in the context of claims. Consequently, it will become necessary to refer to the contract conditions and in some cases to include quotations to illustrate examples of wording and how this may be effectively incorporated within the submission. Rather than confuse the issue by using different examples from different forms of contract, I have made reference throughout the book to the *Conditions of Contract for Construction for Building and Engineering Works Designed by the Employer, First Edition 1999*, published by the Fédération Internationale des Ingénieurs-Conseils (more commonly known as FIDIC), who have kindly given me permission to quote from their publication. I feel that this publication is suitable firstly, because this book is aimed at the international market in which this form of contract is widely used and secondly, because it contains most of the principles included in other forms of contract in one form or another, and the examples used may therefore be easily adapted to suit alternative forms of contract.

It is anticipated that readers will use this book as a guide and as reference material, and thus pick and choose the sections and references appropriate to their particular requirements. It is with this in mind that in some cases, in order to avoid the reader having to refer to other sections to obtain the relevant references, I have repeated the same contract conditions under different subjects.

Definitions

It is necessary throughout the book to refer to various parties, and to describe various actions, events or things. Different forms of contract in widespread use often employ different terminology and, in order to maintain consistency, I have used the same definitions throughout. Most of the terms and/or definitions are consistent with the FIDIC form of contract detailed above but for the sake of clarity they are defined as follows:

'The Employer':   The party who has ordered the work to be done. In other forms of contract, this party may be referred to as 'the Client' or 'the Owner'.

'The Contractor':   The party who is responsible for completing the work.

'The Engineer':   The party responsible for administering the Contract. In other forms of contract, this party may be referred to as 'the Architect', 'the Supervising Officer' or 'the Project Manager'.

'The Reviewer':   The person responsible for reviewing a claim or response and producing a determination, defence or response to the submitted document.

'The Contract':   The various documents including the letter of award, the agreement, contract conditions, the specifications, the drawings and any further document listed in the agreement or letter of award which together form the contract between the parties.

'Programme':   Programme means a breakdown of the work required into a list of planned activities showing the times and dates when they are intended to happen or be done, in the form of a bar chart. In some parts of the world, this is customarily referred to as a 'schedule'.

'Delay Damages':   A sum of money, which shall be paid by the Contractor to the Employer in the case of failure to complete the Project by the Time for Completion. Some contracts deduct such monies under provisions of 'liquidated damages' or 'penalties'.

Currency: Despite the international target market of this publication, where it is necessary to refer to monetary values, I have used Pounds Sterling (£) as the currency. While attempts have been made to make the values fairly realistic, particularly in relationship to each other, any monetary values are purely used for illustrative purposes and make no attempt to reflect true or actual values.

The Example Projects

In order to attempt to bring about a realistic treatment of the example claim and response documents discussed in the forthcoming chapters, I have created a couple of fictional projects and used events which could have conceivably occurred on such projects to demonstrate how a typical claim or response could be written. The examples used are based upon actual situations for which I have prepared either the claim or the determination.

Chapter 2

Contract Administration for Claims and Claims Avoidance

Introduction

In this chapter, we will discuss how good contract administration systems and procedures will help in all matters relating to claims. Contract administration is a far-reaching topic and much of it falls outside the purpose of this publication, but there are many ways in which good contract administration will help to produce a robust and well-substantiated claim and also make life less difficult for the person who has the task of preparing the claim. While we are looking at this from the point of view of the claimant, the party responsible for responding to claims will also benefit greatly from efficient contract administration for exactly the same reasons. Many claims arise from such things as conflicts or ambiguities in contract documents or the failure of one of the parties to comply with their obligations under the contract. Good contract administration can also therefore play an important part in avoiding claims.

The Contract Documents

The contract documents will usually form the basis of any claim. If something has changed, the extent of the change may only be measured and evaluated by reference to the drawings and specification upon which the contract is based. The claimant's entitlement to make a claim will usually be spelled out in the conditions of contract, as will the procedure to be followed in the event of a claim. If there is a disagreement between the parties as to design, quality, responsibility, scope of works or procedures, the contract is the place to look for guidance and resolution.

It should be remembered that while the tender documents should contain the contract conditions, they are not the contract documents. Tender documents address the tenderer and contract documents address the Contractor. While many of the tender documents may be

Construction Claims & Responses: Effective Writing & Presentation, Second Edition. Andy Hewitt.
© 2016 John Wiley & Sons, Ltd. Published 2016 by John Wiley & Sons, Ltd.

Chapter 2

subsequently incorporated into the contract, care must be taken to ensure that they actually belong there.

The contract documents will typically comprise the following:

1. The agreement
2. The particular conditions of contract
3. The general conditions of contract
4. The specification
5. The drawings
6. The bills of quantities
7. Other documents

Most contracts will provide that all the documents shall be mutually explanatory, but in the case of ambiguity or discrepancy, they are to be read in a stated order of precedence. The list above is fairly typical of an order of precedence. (See FIDIC 1999 Red Book, Sub-Clause 1.5 [*Priority of Documents*]).

With regard to the last item on the list 'Other documents', there is a great temptation for those whose job it is to prepare and compile the contract documents, to 'dump' all sorts of documents into this section. Typically these may consist of correspondence between the parties between the time of tender and the letter of acceptance, tender queries and clarifications, tender bulletins, minutes of meetings during the negotiation process, the Contractor's proposals for executing the works, offers to complete the project differently from that of the tender documents or value engineering proposals and such like.

The potential for conflict between such documents and the main contract documents becomes high in such a situation and may, according to the order of precedence stated in the contract, not reflect the pre-contract negotiations and a subsequent agreement between the parties. If we look at a situation whereby the Contractor has negotiated a reduction in the time for interim payments to be made from 56 to 40 days and this was recorded in meeting minutes included in 'Other documents' then, on a strictly contractual basis, the 56 days in the general conditions would still have precedence over such an agreement and effectively render the agreement null and void. The possibility for error is also increased if important points are 'hidden' away in the back of the contract documents. While a good contracts administrator should be aware of potential hazards, would a busy site engineer ever think to look through such a section to check whether a specification item had been changed by way of a tender bulletin, or recorded in a letter or meeting minutes? It is unlikely.

Disputes often arise through the interpretation of the contract, and it is true to say that if the contract documents are poorly drafted and compiled, the potential for disputes increases tremendously. For these reasons it is infinitely preferable to keep the 'Other documents' section as small as possible and to amend the tender documents to take into account any changes that have been negotiated and agreed between the parties within the appropriate section of the contract documents. This also applies to tender queries and their responses, which will arise in the first instance from lack of clarity, ambiguity or conflict within the documents. Rather than just including the tender queries and responses as an addendum to the contract, as is often the practice, the contract documents should be amended to reflect the instructions given in the responses to such queries.

The best time to complete and sign the contract is as soon as possible after the agreement has been made. If this is not done, personnel responsible for the construction and administration of the project will often replace the people involved in the tender, and subsequent negotiations or memories will grow dim. Worse than this, circumstances could arise on site, which would make the inclusion or not of a particular item of negotiation extremely important, a situation that could encourage people to acquire a 'selective memory' of the pre-contract negotiations. In any case, the site personnel need to have an accurate set of documents to tell them what they are going to be building and how they are supposed to build it. While there should be pressure to produce the contract documents for completion and signature as soon as possible, it should always be remembered that rushing this very important task and producing a poor set of contract documents could have serious consequences later on.

It is essential that each of the parties and the party responsible for administering the contract on behalf of the Employer have a full set of contract documents for reference. A good contract administration system would ensure that a controlled set of the contract documents is kept on site and each individual page or sheet of the documents is clearly marked to show that it is a contract document. This last requirement is useful to ensure that when sections of the documents are copied at a later date and when revised drawings and specifications have been issued, these is no confusion with regard to the status of each document.

If there *are* documents included within the 'Other documents' section, then it is a good idea to annotate the controlled copy of the contract documents to draw attention to the fact. For example, a specification clause could be annotated to refer to a tender query and response within the contract documents that includes a change or clarification to the particular specification item. Better still, the query and response could be photocopied and pasted into the specification at the appropriate place. This will ensure that the project team who need to refer to the contract documents will be aware of the true meaning of the contract, and mistakes and wasted effort will be eliminated.

It should go without saying that a person responsible for dealing with claims should have accurate information available as to the provisions of the contract, because this will be the very basis of almost all claims. Anyone dealing with contractual matters will inevitably deal with the conditions of contract on a regular basis. It is therefore good practice to prepare a working document that consolidates the particular conditions and the general conditions of contract. This should consist of a copy of the general conditions of contract to which the particular conditions of contract have been incorporated, either electronically or manually, on a clause-by-clause basis. This will ensure that no misinterpretations will occur through someone forgetting to check if the particular conditions have amended a clause that needs to be referred to or relied upon.

The people responsible for administering the contract and anyone involved in the preparation of contractual letters and claims will have cause to quote from the contract conditions or to reproduce clauses for one reason or another on a regular basis. If a 'soft' copy of the contract is available, then this task becomes quite easy, and provided that the particular conditions of contract have been incorporated, this will prove to be a useful document throughout the life of the project, particularly if it becomes necessary to prepare or to respond to claims. If an editable copy is not available, it is a good idea, each time you need to quote from a clause, to first type the full clause into a separate Word document and thereby gradually build up a 'library' of clauses that may be used to 'cut and paste' from at a later time.

Finally, a constant point of contention when agreeing new rates and prices is the level of the Contractor's overheads and profit to be added to the base cost. It is suggested that the Contractor should be requested to submit a percentage with his tender and then that, providing the rate is acceptable, should be subsequently included in the contract. The Contractor's subcontractors will also require their own overheads and profit to be included in new rates and prices for their works, so if the percentage were extended to the major subcontract trades, then the potential for future conflict would be greatly reduced. If no such provision is included within the contract, it is suggested that an early agreement on these matters should be sought in order to enable the evaluation of new items to be agreed as the work proceeds.

Programmes and Planning

Most forms of contract require the Contractor to submit a programme of the works to the Engineer for acceptance or approval, and this usually has to be done within a specified time frame – the FIDIC Red Book Sub-Clause 8.3 [*Programme*] for example stipulates 28 days. On a complicated project, however, it is a challenge to produce a detailed programme, which may contain thousands of activities within such a time frame. Nevertheless, the Contractor should do his utmost to produce the most detailed and accurate programme possible and to do it within the time stipulated. As a minimum the programme should include:

1. a clear intention of how the time and sequence of how the work is intended to progress
2. a clear critical path to completion
3. the dates when the employer's input is required, with links to the critical path.

If delays occur and the Contractor wishes to pursue an extension of time claim then the programme will be the yardstick against which to measure the effect of delays, so it is essential to have this in place as soon as possible. There seems to be a propensity among engineers to delay acceptance of the Contractor's programme, and in some cases to attempt to pressure the Contractor to change his intentions for the execution of the works, or to allow for events that have happened post-contract. Examples of this would be if the contract period was to start on a particular date, but the Contractor did not receive possession of the site on that date, or a delay to a nomination date. It should be remembered that the programme should be based upon the works included in the Contract and, consequently, in such a situation the Contractor should resist any pressure to change his programme to reflect the actual date of possession or a revised nomination date, because this would not reflect the agreed situation at the time the Contract was entered into. It is usual for contracts to allow for revised programmes to be prepared should the previous version no longer reflect the intended sequencing of the project, current progress or extensions of time awards. The revised programme is the one in which to make amendments and not in the original 'baseline' programme.

Contract administration procedures should allow for an updated programme to be maintained, recorded, submitted, agreed and kept as a record on a regular basis. An updated programme will record the as-built or actual progress up to the data date of the update and will predict the future events, including changes to the critical path and a predicted completion date. The updates should include for changes such as additional work, omitted work and extensions of time awards. If logic errors have been discovered within the baseline programme, then these should be discussed with the Engineer and corrected accordingly. Should it be necessary in a claim situation, to demonstrate whether or not the project was proceeding as planned or that there were no concurrent delays at any time, such a record will become an essential tool in doing so. While updated programmes are often finalised and recorded on a regular, possibly monthly, basis, if an event occurs which may have a significant impact on time, it would be an appropriate measure to create and record the updated situation at the time of the onset of the delay. This would go a long way to proving or disproving entitlement to extensions of time and additional payment when a claim is subsequently submitted.

Many extension of time claims flounder on the method of delay analysis that should be used to demonstrate the effect of delays. The Contractor often wishes to perform one method of delay analysis and the Engineer considers that another method is more appropriate. Such disagreements will only serve to prolong the resolution of a claim and involve the parties in additional expense. It is suggested that if the Employer or Engineer prefers a specific method of delay analysis to be utilised, then this requirement is included within the Contract. This will ensure that the Contractor maintains all the necessary records and produces programmes that will allow an analysis to be performed in the required manner.

If you are responsible for administering, preparing or responding to claims, it is very important to establish a rapport with those responsible for planning and programming and to ensure that they are aware of the requirements with regard to claims. Unless you are proficient in the use of planning software yourself, you will have to work with and depend upon such people when it comes to demonstrating the effects of delay on the programme. Consequently, the early establishment of a good working relationship in this respect is essential.

Contract administration systems should establish procedures for the early identification of potential claims, firstly because notices of such will usually have to be submitted in order to protect entitlement, and secondly so that the contract administration system can 'kick in' and appropriate actions be taken. One of the early things that should be done from a claim point of view is to consult with the planners to ascertain whether an event will have an effect on the programme and, if so, what the likely effect will be. At this stage you are attempting to discover whether a delay to the completion date is likely, and consequently whether a notice of an extension of time should be submitted. An in-depth analysis is not necessary at this stage, but the planners should at least provide a 'best opinion' on which to base the decision as to whether to send a notice and flag the event as a potential claim. It is always better to send a notice and later advise the Engineer that, having investigated the matter further, no claim will be pursued, than not to send a notice.

Many arguments have taken place on the correct allocation of float contained in the programme when it comes to extensions of time. One point of view is that the project owns the float and the other is that because the programme is 'owned' by the Contractor, then any float contained therein also belongs to him. It is suggested that, in order to avoid later differences of opinion on the subject, it is good practice that the conditions of contract stipulate how the float should be used. If the use of float is not defined in the contract, it is sensible for the parties to attempt to agree on the matter and record such an agreement during the early stages of the project.

Records, Records, Records

Many years ago I attended a seminar presented by Roger Knowles and while many of the details have in subsequent years been lost to memory or have hopefully just been absorbed into my everyday knowledge, there is one thing that I remember explicitly. This was when Roger informed the attendees that the three most important aspects to a successful claim are 'good records, good records and good records'. During the ensuing years, I have found absolutely no reason to disagree with this statement and have also come across several situations where a potentially good case has been spoiled for the absence of adequate records.

The burden of the claimant is to prove his case on the balance of probabilities and in order to do so, he must substantiate that the events have actually occurred, possibly substantiate the timing of the events and substantiate that the provisions of the Contract have been complied with in terms of notices and submissions. The only way this may be done is by reference to the project records, and if the claimant does not have such records, it will be very difficult to prove the claim.

Records may comprise letters, memos, transmittals, meeting minutes, daily, weekly and monthly reports, programmes and notices. Important contemporaneous supporting evidence to any claim could include the following:

- Tender and Contract documents
- Daily and/or weekly site-progress data
- Periodic progress reports
- Daily staff, labour and plant records
- Material deliveries to site
- Drawing register showing issue dates and revision numbers
- Contract programme
- As-built programme
- Progress records to show activities started, in progress, on hold, suspended or completed
- Variation Orders or the like

Chapter 2

- Correspondence
- Meeting minutes
- Notices
- Site diaries
- Site memos and instructions
- Photographs with date records
- Site measurement records
- Daywork records (whether or not this will be the eventual means of evaluating additional works)
- Purchase orders and invoices

Many of the above may be submitted to the Engineer (or his representative or equivalent under various forms of contract) on a periodic basis and if such a person can be persuaded to acknowledge them as being accurate contemporaneous records, it would be difficult to repudiate them if they were later relied on as evidence to support a claim. While it may not always be possible to persuade an Employer's agent to 'sign off' on such records, if they are submitted to such a party and are not refuted within a reasonable time, then it would be difficult to deny their accuracy at a later date.

FIDIC Sub-Clause 20.1 [*Contractor's Claims*] includes specific requirements to keep records relating to claim events as follows:

> The Contractor shall keep contemporary records as may be necessary to substantiate any claim, either on the Site or at another location acceptable to the Engineer. Without admitting the Employer's liability, the Engineer may, after receiving any notice under this Sub-Clause, monitor the record-keeping and/or instruct the Contractor to keep further contemporary records. The Contractor shall permit the Engineer to inspect all these records, and shall (if instructed) submit copies to the Engineer.

It is therefore extremely important that, firstly, a robust contract-administration system is created and secondly that systems and procedures are put in place to adequately record the events and the effects on a contemporaneous basis.

The success of a claim will depend on the claimant being able to prove to a reasonable extent the cause and effect of the claimable event; to do this, good record keeping is absolutely essential. Most contract administration systems will allow for events to be recorded by way of correspondence and meeting minutes, and much of the necessary substantiation of contemporaneous events can be obtained from such records. Submittal logs of shop drawings, material approval requests, method statements and the like, which record each action by the various parties are also a useful source of information and record, as are daily site reports that record resources deployed, materials delivered and the activities being worked on. All of these will help to establish dates and times, which, particularly in the case of an extension of time claim, will prove to be of great benefit. Computerised document management systems are common on construction projects and may be used to great effect in claim situations.

The contract administration system should also ensure that additional records are kept when a likely claim event occurs. For example, if the Contractor receives an instruction which requires executed work to be altered or demolished, it would be a good idea to photograph the progress at the time that the instruction was received and to annotate drawings to record the same. It would also be sensible to keep daywork records of time and materials deployed for the alteration or demolition work. Obviously each situation will be different, but it is always better to have kept records that may not eventually be required, than to have no means of proving the events.

Chronology and Database

When it comes to preparing the claim, much time can be saved if various records relating to the matter are collected and recorded on a database. The time to start doing this is as soon as an event occurs which is likely to result in a claim. It then becomes a fairly easy job when correspondence, minutes and so on are reviewed on a daily basis, to copy and file relevant documents separately and to add them to the database. An example of a simple database that may be compiled in a spreadsheet is shown in Table 2.1.

One of the best ways to write a claim narrative, particularly in the case of an extension of time claim, is by way of a chronology of events. This will help to demonstrate cause and effect and will also 'tell the story' in a logical way that automatically leads the reviewer to a logical conclusion. When sorted into chronological order, the records in such a database will form an excellent starting point from which to start the claim narrative.

Table 2.1 Example of a simple database

Johnson Construction Group
Northern ring road, Newtown
Document database – Waterproofing claim

Date	Type	Ref.	From	To	Subject	Contents
08-Jul-13	Letter	0306	JCG	APS	Supply and Apply Eliminator System (Spray Type Waterproofing)	Letter of intent
18-Jul-13	Mat App	44	JCG	ENGR	Materials Approval	Product and subcontractor approval
25-Sep-14	Test	C1194	APS	JCG	Test Certificate	Test Certificate 2A Underpass Bridge
14-Oct-14	Letter	2053	ENGR	JCG	Laying of Sub-grade at Different Locations in Type 1 waterproofing	Waterproofing under investigation. Instructed not to lay until approved. APS to provide requested information
16-Oct-14	Letter	001	JCG	APS	Waterproofing Type 1	Request technical clarifications. Enclosing mix designs
17-Oct-14	Letter	2196	JCG	ENGR	Laying of Sub-grade at Different Locations Waterproofing Type 1	Conf of verbal instructions. Proceed with sub-grade where waterproofing completed
17-Oct-14	Letter	2061	ENGR	JCG	Laying of Sub-grade at 1C-2A on Waterproofing Membrane Type 1	Letter of undertaking required from APS
25-Oct-14	Email	1194	APS	JCG	Waterproofing Type 1	Re Meeting 12 Oct. required RSA mix design to confirm compatibility. Pointing out conflicts in dwgs and request clarification
01-Nov-14	Letter	2114	ENGR	JCG	Waterproofing Membrane Type 1	Repeat request for APS to provide undertaking of product performance
02-Nov-14	Letter	2264	JCG	ENGR	Waterproofing Type 1–APS report	Request instructions re use of RSA and RSA specification
03-Nov-14	Letter	2123	ENGR	JCG	Waterproofing Type 1–APS report	Query why RSA is required. Repeat request for performance undertaking

06-Nov-14	Letter	2280	JCG	ENGR	Waterproofing Type 1–APS report	Repeat request instructions re use of RSA and RSA specification
06-Nov-14	Letter	2136	ENGR	JCG	Waterproofing Type 1–APS report	Request information on use of RSA and performance undertaking. Confirm spec for RSA provided by email
11-Dec-14	Letter	2239	ENGR	JCG	Notification of Claim for Deletion of Waterproofing Membrane	Not following claim procedure
19-Dec-14	Letter	2435	JCG	ENGR	Trough Section Drawings at IC (3)	Not relevant to waterproofing
20-Dec-14	Letter	2438	JCG	ENGR	Notification of Claim for Deletion of Waterproofing Membrane	Following claim procedure. Re-confirm notice of claim
20-Dec-14	Letter	2439	JCG	ENGR	Waterproofing Type 1 for Decks	Confirm instructions deleting waterproofing. Notice of claim
04-Jan-15	Letter	2496	JCG	ENGR	Notification of Claim for Deletion of Waterproofing Membrane	Reserve rights to claim
04-Jan-15	Letter	2497	JCG	ENGR	Notification of Claim for Deletion of Waterproofing Membrane	Notice of claim
05-Jan-15	Letter	2299	ENGR	JCG	Notification of Claim for Deletion of Waterproofing Membrane	Rejection of claim. Allege delivered mats not notified previously
05-Jan-15	Letter	2300	ENGR	JCG	Deletion of Waterproof Trough Section! C-03 Completion	Rejection of delay claim

Such records and the database will be invaluable information to a claims consultant, or to someone who is not based on the project and who is brought in to prepare the claim. The database can also be used as an essential working tool while preparing the claim. For example, significant events may be highlighted to establish when a certain event occurred, and the 'sort' and 'filter' functions in Microsoft Excel may be used to good effect.

Notices

Many forms of contract include requirements for the Contractor to provide notices of events or circumstances that the Contractor considers may provide entitlement to additional time or payment within a specified time frame of the event occurring. The reasoning behind such requirements is that the Employer and his agents need to be made aware of the circumstances as soon as possible, in order that they may consider corrective action or mitigation measures which may be implemented to minimise the effects of the circumstances. Additionally, if no mitigation measures may be undertaken, the prudent employer will include provisions for the forthcoming claim in his budget. It is fair to say that it is probably easier to extract money from a party who has made some sort of financial provision for the event or circumstance, than from one where no such provision exists.

In the author's experience, many contractors for some reason view the obligation to submit notices as being either unnecessary, too much trouble or an obstacle to good relations on a project. If, however, we look at a simple example from everyday life, we can possibly see things from the other side of the fence. Let us imagine that a person has unintentionally defaulted on a car finance repayment. Now, from that person's point of view, would it be reasonable for the finance company to send a tow-truck around to repossess the car without prior notice, or would it be more reasonable to write to the person, enclosing a copy of the loan repayment account, draw their attention to the fact that the account is in arrears and inform them that unless it was brought up to date within a certain time, the car would be repossessed? Most people would surely agree that the second option would be the most reasonable way to deal with such a circumstance. Surely then, it is reasonable to inform the Employer of a circumstance which may lead to a claim.

The Contract will sometimes state the requirements for a notice, will often prescribe time frames within which notices must be submitted and will usually provide details of how notices should be sent or delivered.

The last point is often overlooked by the Parties. The FIDIC Red Book contains the following under Sub-Clause 1.3 [*Communications*]:

Wherever these Conditions provide for the giving or issuing of approvals, certificates, consents, determinations, notices and requests, these communications shall be:

(a) in writing and delivered by hand (against receipt), sent by mail or courier, or transmitted using any of the agreed systems of electronic transmission as stated in the Appendix to Tender; and

(b) delivered, sent or transmitted to the address for the recipient's communications as stated in the Appendix to Tender...

Sub-Clause 20.1 [*Contractor's Claims*] from FIDIC has this to say on the subject of notices:

If the Contractor considers himself to be entitled to any extension of the Time for Completion and/or any additional payment, under any Clause of these Conditions or otherwise in connection with the Contract, the Contractor shall give notice to the Engineer, describing the event or circumstance giving rise to the claim. The notice shall be given as soon as practicable, and not later than 28 days after the Contractor became aware, or should have become aware, of the event or circumstance.

Sub-Clause 20.1 [*Contractor's Claims*] carries on to emphasise the importance of submitting such notices as follows:

If the Contractor fails to give notice of a claim within such period of 28 days, the Time for Completion shall not be extended, the Contractor shall not be entitled to additional payment, and the Employer shall be discharged from all liability in connection with the claim. Otherwise, the following provisions of this Sub-Clause shall apply.

The above provision is a condition precedent to entitlement, which means that if the Contractor does not comply with the requirements of the Contract in this respect, he will lose the rights for compensation to which he would otherwise have been entitled. While there may be legal arguments to counter such a provision in certain jurisdictions, it is surely better to ensure that the contractual provisions for the submission of notices and particulars are complied with, than to be obliged to resort to the law to prove such a point. It is fairly obvious therefore that the Contractor's contract administration systems should place particular importance on the subject of submitting notices in this respect. It is a simple matter to create preformatted notices where the 'blanks' may be filled in as and when it becomes necessary to issue them.

If, on subsequent examination, it is found that no effect will be experienced by the event recorded, then it is a simple task to follow up the notice with a confirmation to that effect to record that the issue is closed. On the other hand, if the Contractor does not submit notices and particulars, he may very well find that his entitlement has been negated.

A notice need not be a long or detailed document because its sole purpose is to draw the attention of the Engineer to the fact that something has happened that may give the Contractor entitlement to make a claim. It should however do the following:

1. State clearly that it is a notice.
2. Provide details of the event or circumstance and the date of its occurrence.
3. State the contractual or legal provisions, which provide entitlement to the claim.
4. State whether the Contractor intends to claim for an extension of time, additional payment of both.
5. Request the Engineer to advise of any records that he requires to be kept.

It should be remembered that a notice is not a claim, and the temptation to go into detail or to attempt to quantify the claim at this stage is not only unnecessary, but may illicit an unwelcome response which attempts to reject the claim before it has been properly considered, prepared and submitted. Such a situation is a waste of time and effort for both parties and is also likely to lead to ill feeling and animosity.

Dispute Adjudication Boards and the Like

In recent years, the appointment of dispute boards has become increasingly popular. In fact, given the impressive statistics on dispute avoidance published by the Dispute Review Board Foundation and the Dispute Board Federation, one may wonder why all projects do not follow this route. A dispute adjudication board consists of a one-person or three-person tribunal that is appointed jointly by the parties to keep abreast of events on the project and to be able to quickly provide an opinion or a decision on matters concerning the Contract. Unlike arbitrators who are appointed when a dispute has already occurred, the advantage of such a board is that the board members can approach a dispute with prior knowledge of the project and the parties and probably even the issue in question. Additionally, the parties may refer a potential problem, possibly with regard to the interpretation of the Contract, to the board and request an opinion. An opinion provided by a panel who are experts in their field will often serve to prevent a matter from escalating to a point where the parties are unable to agree.

There seems to be a propensity for the parties to delay the appointment of the dispute board until a dispute has actually occurred, This flies in the face of the intention to have a panel of experts who are able to advise the parties and provide decisions in a timely manner on an ongoing basis – in other words, the very things that will help to avoid disputes in the first place. It is therefore suggested that the dispute board is appointed without undue delay.

See Chapter 11 for more details of dispute boards.

Procedures

It is not the purpose of this publication to dwell on the subject of contract administration, but at the very least, the following requirements with regard to claims management should be taken into account in the contract administration procedures:

1. The conditions of contract and other contract documents should be examined to ascertain the procedures in the case of claims. In the case of FIDIC, this would comprise the following:
   a. Giving notice to the Engineer, describing the event or circumstance giving rise to the claim as soon as practicable and not later than 28 days after the Contractor became aware, or should have become aware, of the event or circumstance
   b. Submitting other notices that are required by the Contract
   c. Keeping contemporary records
   d. Submitting a fully detailed claim with supporting particulars within 42 days of the event or circumstance giving rise to the claim
   e. Sending interim updates of the claim if the events giving rise to the claim have a continuing effect

f. Submitting final details of the claim within 28 days of the end of the events giving rise to the claim

g. Creating a register to log events that may give rise to claims, to record the dates by which the provisions of the contract with regard to notices and submissions are to be complied with and to record such documents

h. Examining incoming drawings for changes to the works that may lead to a claim for additional time or payment

i. Examining incoming correspondence for instructions that may give rise to delay or additional payment

j. Examining meeting minutes for instructions that may give rise to delay or additional payment

2. Should the above examinations reveal that an event has occurred that may give rise to a claim, then the system should trigger the procedure for the submission of notices and detailed particulars of the claim.

3. Periodic reports prepared and submitted to the Engineer and/or Employer should include brief descriptions of notices and claims submitted and the current status of the claims.

While the above procedures are discussed with regard to the Contractor's perspective, the Engineer would also be well advised to ensure that his contract administration systems are also well considered, because good administration systems will be necessary when it comes to responding to claims.

Chapter 3

Types of Claim

A claim is simply an assertion of a party's right under the terms of a contract or under law. In the construction industry, this more often than not comes down to a right to either additional time to complete the works or additional payment, and is very often a combination of the two. In claim presentations, however, much confusion seems to result from erroneous assumptions regarding delay, extensions of time and money, and it is therefore important to be aware that delay does not always bring about an extension of time and that an extension of time does not always bring about an award of money.

This chapter examines the more common types of claim and discusses how they may be applicable to a simple event on a typical project. The event we will examine is the division of one large room on a project that is under construction, into two smaller rooms by the addition of a new blockwork partition wall, a new doorway, additional lighting and additional electrical power outlets.

Claims for Variations

Simply put, a variation is something that changes the nature of the project. This may consist of additional or less work, a change in the specification for part of the work, or in some cases a change to the contractual basis of the project such as the contract price or period. In most construction contracts, the Employer is permitted to make such changes, and this is usually administered by his appointed agents such as the Engineer under the FIDIC forms of contract.

Most standard forms of contract place the onus on the Engineer to issue instructions to vary the works and most will specifically prohibit the Contractor from carrying out variations without written instructions.

Construction Claims & Responses: Effective Writing & Presentation, Second Edition. Andy Hewitt.
© 2016 John Wiley & Sons, Ltd. Published 2016 by John Wiley & Sons, Ltd.

In an ideal world such changes would be correctly administered by the Engineer by way of formal variation instructions, accompanied by additional information, such as drawings, to enable the Contractor to carry out the varied work and, of course, the instructions would all be issued in plenty of time so as not to disrupt or delay the Contractor's planned sequence of work in any way.

The FIDIC contract has this to say about the subject of variations:

Sub-Clause 1.1.6.9	"Variation" means any change to the Works, which is instructed or approved as a variation order under Clause 13 [*Variations and Adjustments*].
Sub-Clause 13.1 Right to Vary	Variations may be initiated by the Engineer at any time prior to issuing the Taking-Over Certificate for the Works, either by an instruction or by a request for the Contractor to submit a proposal.

The Contractor shall execute and be bound by each variation, unless the Contractor promptly gives notice to the Engineer stating (with supporting particulars) that the Contractor cannot readily obtain the Goods required for the Variation. Upon receiving this notice, the Engineer shall cancel, confirm or vary the instruction.

Each variation may include:

(a) changes to the quantities of any item of work included in the Contract (however, such changes do not necessarily constitute a Variation),
(b) changes to the quality and other characteristics of any item of work,
(c) changes to the levels, positions and/or dimensions of any part of the Works,
(d) omission of any work unless it is to be carried out by others,
(e) any additional work, Plant, Materials or services necessary for the Permanent Works, including any associated Tests on Completion, boreholes and other testing and exploratory work, or
(f) changes to the sequence or timing of the execution of the Works.

The Contractor shall not make any alteration and/or modification of the Permanent Works, unless and until the Engineer instructs or approves the Variation.

| Sub-Clause
13.3 Variation
Procedure | If the Engineer requests a proposal, prior to issuing a Variation, the Contractor shall respond in writing as soon as practicable, either by giving reasons why he cannot comply (if this is the case) or by submitting: |

(a) a description of the proposed work to be performed and a programme for its execution,
(b) the Contractor's proposal for any necessary modifications to the programme according to Sub-Clause 8.3 [*Programme*] and to the Time for Completion, and
(c) the Contractor's proposal for evaluation of the Variation.

The Engineer shall, as soon as practicable after receiving such proposal (under Sub-Clause 13.2 [*Value Engineering*] or otherwise), respond with approval, disapproval or comments. The Contractor shall not delay any work while awaiting a response.

Each instruction to execute a Variation, with any requirements for the recording of Costs, shall be issued by the Engineer to the Contractor, who shall acknowledge receipt.

Each variation shall be evaluated in accordance with Clause 12 [*Measurement and Evaluation*], unless the Engineer instructs or approves otherwise in accordance with This Clause.

The above provisions are quite clear as to the authority of the Engineer to vary the Works and the procedures to be adopted should a variation be contemplated or instructed. If this procedure is adhered to, there would be no reason to submit a claim because the variation and entitlement to additional payment have already been acknowledged.

In reality, however, the Contractor may not be requested to submit a proposal, and instructions for variations may not be issued under a formal variation order. It is quite common for instructions to be issued by letter, by a response to a request for information, by the issue of revised drawings or by way of a verbal instruction. In many cases the instruction, however it is issued, will not acknowledge the fact that it comprises a variation to the works. Sub-Clause 3.3 of FIDIC contemplates that this may be the case and provides the following:

Sub-Clause 3.3 Instructions of the Engineer	The Engineer may issue to the Contractor (at any time) instructions and additional or modified Drawings which may be necessary for the execution of the Works and the remedying of any defects, all in accordance with the Contract. The Contractor shall only take instructions from the Engineer, or from an assistant to whom the appropriate authority has been delegated under this Clause. If an instruction constitutes a variation, Clause 13 [*Variations and Adjustments*] shall apply.

How does the Contractor therefore ensure that he is compensated properly in a case where he considers that he has an entitlement to additional payment for an instruction which he considers comprises a variation, but which has not been instructed or approved as a variation order?

Firstly, the Contractor should acknowledge receipt of the instruction and confirm that he considers the instruction to be a variation under the provisions of Sub-Clause 13.3 [*Variations and Adjustments*]. The Engineer may agree with this and, if so, he should issue a variation order as confirmation. If no such confirmation is forthcoming or if the Engineer disagrees, then the Contractor's recourse under FIDIC is contained in the provisions of Clause 20 [*Claim, Disputes and Arbitration*], which has this to say:

Sub-Clause 20.1 Contractor's Claims	If the Contractor considers himself to be entitled to any extension of the Time for Completion and/or any additional payment, under any Clause of these Conditions or otherwise in connection with the Contract, the Contractor shall give notice to the Engineer, describing the event or circumstance giving rise to the claim. The notice shall be given as soon as practicable, and not later than 28 days after the Contractor became aware, or should have become aware, of the event or circumstance. If the Contractor fails to give notice of a claim within such period of 28 days, the Time for Completion shall not be extended, the Contractor shall not be entitled to additional payment, and the Employer shall be discharged from all liability in connection with the claim. Otherwise, the following provisions of this Sub-Clause shall apply.

The Contractor shall also submit any other notices which are required by the Contract, and supporting particulars for the claim, all as relevant to such event or circumstance.

The Contractor shall keep contemporary records as may be necessary to substantiate any claim, either on the Site or at another location acceptable to the Engineer. Without admitting the Employer's liability, the Engineer may, after receiving any notice under this Sub-Clause, monitor the record-keeping and/or instruct the Contractor to keep further contemporary records. The Contractor shall permit the Engineer to inspect all these records, and shall (if instructed) submit copies to the Engineer.

Within 42 days after the Contractor became aware (or should have become aware) of the event or circumstances giving rise to the claim, or within such period as may be proposed by the Contractor and approved by the Engineer, the Contractor shall send to the Engineer a fully detailed claim which includes full supporting particulars of the basis of the claim and of the extension of time and/or additional payment claimed. If the event or circumstance giving rise to the claim has a continuing effect:

(a) this fully detailed claim shall be considered as interim;
(b) the Contractor shall send further interim claims at monthly intervals, giving the accumulated delay and/or amount claimed, and such further particulars as the Engineer may reasonably require; and
(c) the Contractor shall send a final claim within 28 days after the end of the effects resulting from the event or circumstance, or within such other period as may be proposed by the Contractor and approved by the Engineer.

Chapter 3

The remainder of this sub-clause deals with the Engineer's obligations and the result of the Contractor's failure to comply with the provisions of the sub-clause.

The above sub-clause thus provides that should the Contractor consider that an instruction given by the Engineer comprises a variation for which a variation order should be issued, then he should give notice of the event or circumstances to the Engineer and proceed to submit a fully detailed claim. Both these actions should occur within the prescribed time frames. The submission of such a notice may very well result in the issue of a variation order, in which case the claim would cease to be relevant as entitlement to any additional payment would be provided under Clause 13 [*Variations and Adjustments*]. It would, however, be prudent of the Contractor not to assume that a variation order will be forthcoming and to proceed with the submission of a claim within the prescribed time frame in order not to find himself having prevented or prejudiced the Engineer from making a proper investigation into the circumstances.

In our example of the division of a room into two rooms, provided that the instruction for the revised design was issued in such time that the Contractor could incorporate the work into his construction activities with no significant effect on his programme, the claim would consist solely of the additional work to provide foundations, blockwork, plasterwork, skirtings, decoration work, the new door, lighting and electrical power outlets. The claim could be measured and evaluated in the manner prescribed in the Contract, which is usually at the rates and prices contained therein.

Such a claim for a variation should be presented in accordance with the principles discussed elsewhere in this book.

Claims for Extensions of Time

Other than variations, the most common reason for the submission of a claim is for an extension to the time for completion of the works. The reasons for such a claim are twofold. It establishes a new completion date for the project and also prevents the Engineer from deducting liquidated damages or penalties for delay, which he is obliged to do under most contracts. An award of an extension of time may also, however, entitle the Contractor to additional payment as compensation for being obliged to maintain his site establishment and contribute to other overheads for a longer period than was contemplated within the Contract. The latter case is referred to as a claim for prolongation costs.

Most forms of contract include specific provisions with regard to extensions of time, the circumstances under which they may be granted and the procedures to be followed by the parties in the event of an extension of time being warranted. FIDIC is no exception and has this to say on the matter:

Sub-Clause 8.4 Extension of Time for Completion	The Contractor shall be entitled subject to Sub-Clause 20.1 [*Contractor's Claims*] to an extension of the Time for Completion if and to the extent that completion for the purposes of Sub-Clause 10.1 [*Taking Over of the Works and Sections*] is or will be delayed by any of the following causes:

(a) a Variation (Unless an adjustment to the Time for Completion has been agreed under Sub-Clause 13.3 [*Variation Procedure*] or other substantial change in the quantity of an item of work included in the Contract,

(b) a cause of delay giving an entitlement to extension of time under a Sub-Clause of these Conditions,

(c) exceptionally adverse climatic conditions,

(d) Unforeseeable shortages in the availability of personnel or Goods caused by epidemic or governmental actions, or

(e) Any delay, impediment or prevention caused by or attributable to the Employer, the Employer's personnel, or the Employer's other contractors on the Site.

If the Contractor considers himself to be entitled to an extension of the Time for Completion, the Contractor shall give notice to the Engineer in accordance with Sub-Clause 20.1 [*Contractor's Claims*]. When determining each extension of time under Sub-Clause 20.1, the Engineer shall review previous determinations and may increase, but shall not decrease, the total extension of time.

Chapter 3

Thus, the above provisions actually *require* the Contractor to submit a claim in order to start the procedure by which an extension of time may be awarded. Sub-Clause 20.1 [*Contractor's Claims*] outlines the following requirements for the submission of claims:

Sub-Clause 20.1 Contractor's Claims	If the Contractor considers himself to be entitled to any extension of the Time for Completion and/or any additional payment, under any Clause of these Conditions or otherwise in connection with the Contract, the Contractor shall give notice to the Engineer, describing the event or circumstance giving rise to the claim. The notice shall be given as soon as practicable, and not later than 28 days after the Contractor became aware, or should have become aware, of the event or circumstance.
	If the Contractor fails to give notice of a claim within such period of 28 days, the Time for Completion shall not be extended, the Contractor shall not be entitled to additional payment, and the Employer shall be discharged from all liability in connection with the claim. Otherwise, the following provisions of this Sub-Clause shall apply.

The Contractor shall also submit any other notices which are required by the Contract, and supporting particulars for the claim, all as relevant to such event or circumstance.

The Contractor shall keep contemporary records as may be necessary to substantiate any claim, either on the Site or at another location acceptable to the Engineer. Without admitting the Employer's liability, the Engineer may, after receiving any notice under this Sub-Clause, monitor the record-keeping and/or instruct the Contractor to keep further contemporary records. The Contractor shall permit the Engineer to inspect all these records, and shall (if instructed) submit copies to the Engineer.

Within 42 days after the Contractor became aware (or should have become aware) of the event or circumstances giving rise to the claim, or within such period as may be proposed by the Contractor and approved by the Engineer, the Contractor shall send to the Engineer a fully detailed claim which includes full supporting particulars of the basis of the claim and of the extension of time and/or additional payment claimed. If the event or circumstance giving rise to the claim has a continuing effect:

(a) this fully detailed claim shall be considered as interim;
(b) the Contractor shall send further interim claims at monthly intervals, giving the accumulated delay and/or amount claimed, and such further particulars as the Engineer may reasonably require; and
(c) the Contractor shall send a final claim within 28 days after the end of the effects resulting from the event or circumstance, or within such other period as may be proposed by the Contractor and approved by the Engineer.

The remainder of this sub-clause deals with the Engineer's obligations and the result of the Contractor's failure to comply with the provisions of the sub-clause.

If we examine the variation discussed previously, if the instruction to divide the room was given at a time when the project was in the final stages before handover, then this would have a significant effect on the Contractor's ability to complete the project on time, not because of the *amount* of additional work, but because of the *timing* of the instruction. If this were the case then, in addition to the claim for the additional work, the Contractor would be entitled to make a claim for an extension of time.

Claims for Additional Payment Due to Prolongation

While an award of an extension to the time for completion will negate the Employer's entitlement to deduct liquidated damages or penalties, such an award does not usually automatically carry an entitlement to any additional payment for the prolonged period. Such a payment may, however, be warranted in circumstances whereby the Contractor, due to the circumstances that have caused an extension of time to be awarded, has been obliged to maintain his site establishment and utilise his head-office facilities and the like for a period longer than anticipated. In such a case, the Contractor is usually obliged to submit a claim for the additional payment.

Returning to our example, if the late division of the room into two separate rooms caused the handover of the project to be delayed by two weeks, then the Contractor would be entitled to claim for compensation for his on-site and off-site overheads for the two-week delay period. This would be *in addition* to the payment for the additional work which would be measured and evaluated at the rates and prices contained in the Contract, because his prolongation costs would not otherwise be compensated through such an evaluation.

Again, under FIDIC, the Contractor's recourse is contained in the following provisions:

Sub-Clause 20.1 Contractor's Claims	If the Contractor considers himself to be entitled to any extension of the Time for Completion and/or any additional payment, under any Clause of these Conditions or otherwise in connection with the Contract, the Contractor shall give notice to the Engineer, describing the event or circumstance giving rise to the claim. The notice shall be given as soon as practicable, and not later than 28 days after the Contractor became aware, or should have become aware, of the event or circumstance.
	If the Contractor fails to give notice of a claim within such period of 28 days, the Time for Completion shall not be extended, the Contractor shall not be entitled to additional payment, and the Employer shall be discharged from all liability in connection with the claim. Otherwise, the following provisions of this Sub-Clause shall apply.
	The Contractor shall also submit any other notices which are required by the Contract, and supporting particulars for the claim, all as relevant to such event or circumstance.

The Contractor shall keep contemporary records as may be necessary to substantiate any claim, either on the Site or at another location acceptable to the Engineer. Without admitting the Employer's liability, the Engineer may, after receiving any notice under this Sub-Clause, monitor the record-keeping and/or instruct the Contractor to keep further contemporary records. The Contractor shall permit the Engineer to inspect all these records, and shall (if instructed) submit copies to the Engineer.

Within 42 days after the Contractor became aware (or should have become aware) of the event or circumstances giving rise to the claim, or within such period as may be proposed by the Contractor and approved by the Engineer, the Contractor shall send to the Engineer a fully detailed claim which includes full supporting particulars of the basis of the claim and of the extension of time and/or additional payment claimed. If the event or circumstance giving rise to the claim has a continuing effect:

(a) this fully detailed claim shall be considered as interim;
(b) the Contractor shall send further interim claims at monthly intervals, giving the accumulated delay and/or amount claimed, and such further particulars as the Engineer may reasonably require; and
(c) the Contractor shall send a final claim within 28 days after the end of the effects resulting from the event or circumstance, or within such other period as may be proposed by the Contractor and approved by the Engineer.

The remainder of this sub-clause deals with the Engineer's obligations and the result of the Contractor's failure to comply with the provisions of the sub-clause.

It is common practice for contractors to link claims for prolongation costs to claims for extensions of time and to present both as one single claim. This is logical because both claims usually arise out of the same event. While this may be appropriate in straightforward circumstances, consideration should be given to dealing with the two subjects separately on the basis that the award of the extension of time is invariably easier to agree and determine because it does not (at this stage at least) cost the Employer any money. Thus, a claim for an extension of time alone can often be dealt with relatively quickly, whereas additional payment claims often, by their very nature, become protracted while details of monetary calculations for prolongation are considered, discussed and negotiated. It is also likely that different personnel will examine the time and monetary issues. Additionally, while the Contractor may claim a certain number of days for an extension of time, it could well be that the Engineer will determine that a lesser time is warranted and make an appropriate award. If this is the case, the Contractor's prolongation claim, which will at this point be based on the claimed extended period, will have to be revised and resubmitted in accordance with the Engineer's determination. If it is decided to submit two separate claims, care must be taken to comply with the time frames prescribed in the Contract for the submission of the prolongation costs. A sensible way of dealing with this is to submit both the extension of time claim and the claim for additional payment at the same time, but as separate and discrete submissions.

Acceleration and Disruption Claims

Whole books and many articles have been written on the subject of acceleration and disruption claims – concepts that are notoriously extremely difficult to prove and to calculate. It is not the purpose of this publication to expound on such matters, except to discuss briefly the circumstances whereby such a claim may be brought.

It could be that the Employer has prevailed upon the Contractor to accelerate the work to either mitigate delays caused for which the Employer is allocated the risk, or simply to complete the project earlier than the time for completion provided for in the Contract. In such circumstances, the Contractor will no doubt incur additional costs in doing so and consequently be entitled to additional payment for the following:

- Extended working hours, resulting in additional costs of labour due to overtime payments
- Increased plant and labour, resulting in additional mobilisation and demobilisation costs
- Increased plant and labour, causing loss of efficiency as a result of too many workfaces operating at once
- Loss of production caused by working out of sequence, or working on more than the optimum number of workfaces

Similarly, while acts of the Employer may not always cause delay, they may cause disruption to the Contractor's planned method of working, leading to loss of efficiency or additional costs incurred in order to mitigate the effects of the disruption.

In the case of our example variation whereby the Contractor has been instructed to divide a room into two separate rooms at a very late stage of the programme, the following must be considered:

- The floor would have to be broken out and the foundations laid at a time when decorations had been completed.

Chapter 3

- Existing walls, finishes, electrical works and decorations would require alterations to incorporate the additional work.
- The additional door, frame and ironmongery would have to be specially ordered and possibly manufactured.
- The required labour and plant resources to carry out this work would possibly have to be remobilised.
- Additional protection would have to be provided to protect the completed finishes along access routes to the working area and to the areas adjacent to where the work is to take place.
- Production rates would be very different from those achieved if the additional work were incorporated within the normal construction programme at the same time as similar activities were in progress.

In this case, the Contractor would not be adequately compensated by measurement and evaluation of the additional work at the Contract rates and prices, or by the addition of a further payment for his prolongation costs due to an extension of time. His recourse would be to either claim for a different means of evaluation, or to claim for additional payment due to the fact that the work was unable to be performed at the time when work of a similar nature was programmed.

An acceptable method of demonstrating and calculating the effects of disruption would be to compare the production of similar work during an undisrupted period with that during the disrupted period. This is an excellent example of the necessity of keeping good records, and many a disruption claim has failed because the claimant has not been able to demonstrate via a 'measured mile' approach what his undisrupted production rates actually were.

In the case of FIDIC, the same claim provisions apply as in the previous examples as follows:

Sub-Clause 20.1 Contractor's Claims	If the Contractor considers himself to be entitled to any extension of the Time for Completion and/or any additional payment, under any Clause of these Conditions or otherwise in connection with the Contract, the Contractor shall give notice to the Engineer, describing the event or circumstance giving rise to the claim. The notice shall be given as soon as practicable, and not later than 28 days after the Contractor became aware, or should have become aware, of the event or circumstance.

If the Contractor fails to give notice of a claim within such period of 28 days, the Time for Completion shall not be extended, the Contractor shall not be entitled to additional payment, and the Employer shall be discharged from all liability in connection with the claim. Otherwise, the following provisions of this Sub-Clause shall apply.

The Contractor shall also submit any other notices which are required by the Contract, and supporting particulars for the claim, all as relevant to such event or circumstance.

The Contractor shall keep contemporary records as may be necessary to substantiate any claim, either on the Site or at another location acceptable to the Engineer. Without admitting the Employer's liability, the Engineer may, after receiving any notice under this Sub-Clause, monitor the record-keeping and/or instruct the Contractor to keep further contemporary records. The Contractor shall permit the Engineer to inspect all these records, and shall (if instructed) submit copies to the Engineer.

Within 42 days after the Contractor became aware (or should have become aware) of the event or circumstances giving rise to the claim, or within such period as may be proposed by the Contractor and approved by the Engineer, the Contractor shall send to the Engineer a fully detailed claim which includes full supporting particulars of the basis of the claim and of the extension of time and/or additional payment claimed. If the event or circumstance giving rise to the claim has a continuing effect:

(a) this fully detailed claim shall be considered as interim;
(b) the Contractor shall send further interim claims at monthly intervals, giving the accumulated delay and/or amount claimed, and such further particulars as the Engineer may reasonably require, and
(c) the Contractor shall send a final claim within 28 days after the end of the effects resulting from the event or circumstance, or within such other period as may be proposed by the Contractor and approved by the Engineer.

The remainder of this sub-clause deals with the Engineer's obligations and the result of the Contractor's failure to comply with the provisions of the sub-clause.

Essentially, claims for acceleration or disruption are recoverable, but the matter of proving the quantum of such a claim will be central to the question as to whether or not the claiming party will in fact recover. Records in such a situation will be essential to the success.

Claims for Damages Under Law

Many forms of contract include specific circumstances that entitle the Contractor to make a claim for additional time and/or payment. FIDIC is quite comprehensive about this and the conditions include the following circumstances under which the Contractor may make such a claim:

Sub-Clause 1.9	Delayed Drawings or Instructions
Sub-Clause 2.1	Right of Access to the Site
Sub-Clause 4.7	Setting Out
Sub-Clause 4.12	Unforeseeable Physical Conditions
Sub-Clause 4.24	Fossils
Sub-Clause 7.4	Testing
Sub-Clause 8.4 (a)	Variations
Sub-Clause 8.4 (c)	Exceptionally Adverse Climatic Conditions
Sub-Clause 8.4 (d)	Unforeseeable shortages in the availability of personnel or Goods caused by epidemic or governmental actions
Sub-Clause 8.4 (e)	Any delay, impediment or prevention caused by or attributable to the Employer, the Employer's personnel, or the Employer's other contractors on the Site
Sub-Clause 8.5	Delays Caused by Authorities
Sub-Clause 10.3	Interference with Tests on Completion
Sub-Clause 13.7	Adjustments for Changes in Legislation
Sub-Clause 13.8	Adjustments for Changes in Costs
Sub-Clause 16.1	Contractor's Entitlement to Suspend Work
Sub-Clause 17.4	Consequences of Employer's Risks
Sub-Clause 19.4	Consequences of Force Majeure

Should circumstances arise that fall outside those specified in the Contract, it would be necessary to make a claim for damages under law. There are many excellent publications that expound on the law and provide case studies to illustrate legal precedent that may be of assistance in such a case, but in this author's experience, a great deal of mileage under many circumstances may be obtained from the law of prevention.

Simply put, the prevention principle is that each party has an obligation to positively enable the other to perform his obligations under the Contract and that one party may not take advantage of the other if the non-fulfilment of an obligation has been as a result of a preventative action by himself.

Contracts usually recognise that the Employer and the Engineer must positively cooperate for the good of the project under the provisions that deal with the following:

- Providing possession and access to the site
- The issue of drawings and information in a timely manner
- Providing instructions for the nomination of subcontractors and suppliers in a timely manner
- Coordination and access by the Employer's other contractors
- The serving of notices when specific actions are required by the other party
- The administration of extensions of time and other claims within specified time frames
- Submission of payment applications, issue of payment certificates and payment within specified time frames

Accordingly, it is a fundamental principle of law that there exists an implied provision in every contract that neither party to the contract will do anything to prevent performance thereof by the other party, or commit any act that will hinder or delay such performance. Thus, the Contractor is entitled to be provided with a reasonable opportunity to perform the Contract without obstruction or interference and the Employer is obliged not to do anything that will hinder or delay the Contractor in performance of the Contract. Should the Employer or his agents prevent the Contractor from performing his obligations properly, then the Employer has committed a breach of contract and the Contractor, under many legal jurisdictions, has entitlement under law to adequate compensation. Such compensation should be based upon an amount that would place the claimant in the same position in which he would have been had the breach not occurred.

If we return to our example event and consider a scenario whereby the Employer had moved equipment into the room that was to be divided into two and it was not possible for the Contractor to commence the work immediately on receipt of the instruction, this would be considered as an act of prevention which would delay the start of the additional work until such time as the Employer removed the equipment. This, in turn, would affect both the extension of time and the additional payment for prolongation to which the Contractor would be entitled.

Interim and Final Claims

In some cases the event which gives rise to the claim may not have ended or it may not be possible to ascertain the final effects of the event when the prescribed period for the submission of the claim has

Chapter 3

expired. An example of such may be where a variation instruction includes an item that is subject to special order and the manufacture and delivery periods have not been finalised at the time when the claim is due to be submitted. As with notices, the Employer and his agents need to be made aware of the effects to enable them to plan for the financial and time implications on the project and to continue to consider mitigation measures. Consequently, many forms of contract require the claimant to provide interim information on the predicted effects at regular intervals. Sub-Clause 20.1 [*Contractor's Claims*] from FIDIC stipulates the following in this regard:

'Within 42 days after the Contractor became aware (or should have become aware) of the event or circumstances giving rise to the claim, or within such period as may be proposed by the Contractor and approved by the Engineer, the Contractor shall send to the Engineer a fully detailed claim which includes full supporting particulars of the basis of the claim and of the extension of time and/or additional payment claimed. If the event or circumstance giving rise to the claim has a continuing effect:

(a) this fully detailed claim shall be considered as interim;
(b) the Contractor shall send further interim claims at monthly intervals, giving the accumulated delay and/or amount claimed, and such further particulars as the Engineer may reasonably require; and
(c) the Contractor shall send a final claim within 28 days after the end of the effects resulting from the event or circumstance, or within such other period as may be proposed by the Contractor and approved by the Engineer.'

The above requirements are quite specific in that the Contractor is obliged to send monthly updates until the final effects of the circumstances of the claim can be ascertained. Such interim updates should be based upon the Contractor's best estimate from the information to hand at the time the update is prepared. In such a case, each submission should be clearly designated as an interim or final claim in much the same way as are applications for interim progress payments and final accounts.

If a claim is based on several events, then interim claims should be submitted until all the events in question are closed out with the full effects demonstrated.

Presentation

However hard the reviewer of a claim submission or a claim response endeavours to remain impartial, if the reviewer's job is made difficult, unpleasant, annoying or unnecessarily complicated, it is unlikely that they will remain truly unbiased towards the party who has caused such inconvenience. On the other hand, a well-presented, well-structured and user-friendly document will go a long way to persuading the reviewer that the party responsible for compiling it knows what they are doing. If this is the case, then the reviewer will probably also consider that the document will have some merit even before he carries out an in-depth review. In short, it is very important to do whatever you can to make the reviewer's job as easy and as pleasant as possible in order to gain whatever sympathy it is possible to give. The importance of a clear and logical presentation cannot be overstated. Assuming that the claim is valid, if the presentation is good and the recipient genuinely impartial, then the claimant is starting from a position of strength.

Presentation of the Submission or Review Document

The claim submission should basically consist of two parts:

1. The narrative which deals with the details of the project, sets out the circumstances of the claim, demonstrates the effects of the claim and explains the basis of any supporting documents that have been prepared to help demonstrate the effects and/or quantum of the claim.
2. Appendices which contain documents such as programmes, calculations and project records that have been prepared to support, illustrate or substantiate the claim.

A well-presented document demonstrates the professionalism not only of the company presenting it, but also of the individual(s) responsible for compiling it and will include the following:

- Good-quality files or folders containing the documents
- Covers and spine labels showing the necessary information
- Clearly tabbed and labelled dividers to separate exhibits and appendices
- Documents printed on good-quality paper
- User-friendly and attractive page layouts using good-quality graphics
- Headers and footers showing the name of the party making the submission, the title of the document, page numbers, the document date and, if necessary, the revision number
- Documents included as exhibits presented in a logical order, clearly and easily identifiable and cross-referenced
- The submission contained in separate volumes and/or sections, which are in a logical order, clearly labelled and clearly referenced

Writing Style

It is highly unlikely that the reviewer of a claim or determination document will regard it in the same way as a well-written novel but, as has been discussed previously, it is important to make the document as user-friendly as possible and this includes the writing style. The narrative should flow, be easily readable and, of course, it should be properly understood. It is very annoying for a reviewer to have to reread parts of a narrative in order to try to make sense of what is written. Use of correct grammar and punctuation as well as a free-flowing writing style is important here.

It may be that someone intimately involved on a project would understand that, say, 'BL2' is being used to save writing 'Basement Level 2' each and every time, but an outside party might not understand the meaning without further explanation. For this reason, abbreviations and acronyms should be avoided unless they are in common usage within the industry and would be understood by those not familiar with the project. At the very least, abbreviations and acronyms that are used should be defined the first time that they are used in the document, for example, 'Basement Level 2 (hereinafter referred to as BL2)'. Similarly, the use of legalese, Latin tags, obscure or infrequently used words and complicated language can have an equally detrimental effect on a proper understanding of a narrative. This is particularly relevant if some of the people involved in the issue do not have the language in which the document is written as a first language. Claims 'experts' sometimes have a propensity to bolster up their arguments, or possibly to attempt to impress the reviewer with their knowledge, by the frequent use of complicated language and legalese. While this may possibly have a place if the reviewer is an arbitrator, lawyer or judge whose vocation demands an understanding of such language, it may not have the desired effect on a resident engineer who has spent his professional career on site, among mud, steel and concrete rather than in a courtroom. The use of 'impressive-sounding' language will never replace a well-presented, easily understood and properly substantiated case that deals with the facts and submissions in a logical manner. It is therefore better to use the most simple and direct language possible to provide a proper understanding of the points being made.

Ambiguities can often be created by the use of such words as 'them', 'they', 'him' and 'it' when referring to the parties, organisations or people. Examination of legal and contractual documents will reveal that the parties are usually referred to as 'the Claimant', 'the Respondent', 'the Employer', 'the Contractor', 'the Engineer' or suchlike. This is to ensure that there is absolutely no ambiguity or confusion as to which party is being referred to. This may mean that a little more typing is necessary, but the small amount of extra effort is worth it if confusion and ambiguity are to be avoided and if it means that the reviewer will easily and correctly understand the narrative.

The use of quotations in a narrative is a very useful tool, which adds credibility to and helps to substantiate statements made within the narrative. Extracts of the Contract may be reproduced to good effect, as may extracts from correspondence and other project records. It is important here to identify quotations properly by the use of quotation marks. British usage tends to prefer single quotation marks 'thus' while American usage prefers double quotation marks "thus". Whichever you decide to use, the important thing here is to maintain consistency throughout the document so as not to confuse the reader. When using quotation marks, the words contained between the marks should be the *exact* words of the passage being quoted. If incorrect grammar, punctuation or spelling are included in the passage, then these should also be reproduced in the quotation. If it is necessary to add clarity or an explanation in the middle of a quotation, such an explanation should be included in brackets [thus].

It is sometimes a good idea to reproduce whole clauses from the Contract or other documents within the narrative in order that the reviewer is made aware of the provisions. In certain cases, however, either the clause is complicated and needs further explanation, or certain parts of the clause are irrelevant to the point being made. In such cases, it may be necessary either to offer an explanation in simple language as to the meaning of the clause, or to reduce the clause into something that provides a clearer understanding of the issue in question. If we take FIDIC Sub-Clause 8.4 [*Extension of Time for Completion*] as an example, here is the full reproduction:

Chapter 4

The Contractor shall be entitled subject to Sub-Clause 20.1 [*Contractor's Claims*] to an extension of the Time for Completion if and to the extent that completion for the purposes of Sub-Clause 10.1 [*Taking Over of the Works and Sections*] is or will be delayed by any of the following causes:

(a) a Variation (unless an adjustment to the Time for Completion has been agreed under Sub-Clause 13.3 [*Variation Procedure*]) or other substantial change in the quantity of an item of work included in the Contract,

(b) a cause of delay giving an entitlement to extension of time under a Sub-Clause of these Conditions,

(c) exceptionally adverse climatic conditions,

(d) Unforeseeable shortages in the availability of personnel or Goods caused by epidemic or governmental actions, or

(e) any delay, impediment or prevention caused by or attributable to the Employer, the Employer's Personnel, or the Employer's other contractors on the Site.

If the Contractor considers himself to be entitled to an extension of the Time for Completion, the Contractor shall give notice to the Engineer in accordance with Sub-Clause 20.1 [*Contractor's Claims*]. When determining each extension of time under Sub-Clause 20.1, the Engineer shall review previous determinations and may increase, but shall not decrease, the total extension of time.

Let us assume that the claimant is preparing an extension-of-time claim due to adverse climatic conditions. Firstly, the entire clause is quite a lot for the reviewer to digest and secondly, it contains quite a lot of information that is irrelevant to the specific subject of the climatic conditions. Here is how it could be reproduced in such a manner to deal effectively with the issue in question:

The Contractor shall be entitled ... to an extension of the Time for Completion if and to the extent that completion ... is or will be delayed by ... exceptionally adverse climatic conditions...

If the Contractor considers himself to be entitled to an extension of the Time for Completion, the Contractor shall give notice to the Engineer ... When determining each extension of time ... the Engineer shall review previous determinations and may increase, but shall not decrease, the total extension of time.

Where parts of the clause are omitted from quotations, the reader's attention should be drawn to this by the inclusion of '…' to indicate the missing passage. If adopting this method of dealing with extracts from the Contract, care should be taken that the *meaning* is not changed by such omissions.

Sometimes it is more appropriate to include sections from the clause into the narrative in order to emphasise the meaning or to ensure that the exact wording is reproduced to avoid ambiguity. Here is an example based upon the same sub-clause from FIDIC:

Sub-Clause 8.4 [*Extension of Time for Completion*] provides that 'the Contractor shall be entitled' to an extension of time if the completion is delayed due to 'exceptionally adverse climatic conditions'. If such an event occurs, the Contractor is obliged to 'give notice to the Engineer in accordance with Sub-Clause 20.1 [*Contractor's Claims*]' and the Engineer is obliged to determine the extension of time.

The above example is succinct, easily understood, carries the authority of the contract and is unambiguous.

The author of the narrative obviously has detailed knowledge of the project and the circumstances surrounding the issues that are the subject of the claim; in fact, he may have lived and breathed the project for many months. It is for this reason that such a person may easily assume, even inadvertently, that a reviewer has the same depth of knowledge as himself. Consequently, and even with the best intentions, the author may fail to make things crystal clear and miss out minor but important explanatory details. When the narrative is complete it is therefore good practice to have someone review the document, and if this person has no knowledge of the circumstances, then that is a distinct advantage. The reviewer should put themselves in the place of the person who will eventually have the task of reviewing the document and advise the author on unclear passages, incorrect grammar, unsubstantiated statements and the like. The reviewer should also refer to any programmes, calculations and so on that are referenced in the document to ensure that the narrative has incorporated the correct information, that explanations contained in the narrative are easily followed and that any cross-references to other documents are correct. It should also go without saying that any calculations should be mathematically checked. It may be tempting to proofread a document that you have prepared yourself. This is not good practice, as it is normal when proofreading your own work, to read what you *think* that you have written and not what you have *actually* written. A review by another person will usually pick up mistakes that you would not have identified when proofreading your own work.

Chapter 4

Key Points for Claim Presentation

There are four key points to bear in mind when preparing a claim document. If these are followed they will go a long way towards helping a just claim succeed in minimum time.

Key Point No. 1 – Make the reader's job as easy and as pleasant as possible.

Key Point No. 2 – Ensure that the submission is a stand-alone document.

Key Point No. 3 – Assume that the reviewer has no prior knowledge of the project or the circumstances.

Key Point No. 4 – Don't include irrelevant information in the document.

The above are discussed in the following sections.

Making the Document User-Friendly

The document submitted will become a working document for the person responsible for reviewing it. Seemingly small things will make this person's job much more pleasant, or at any rate may prevent him becoming antagonistic towards you if you make his job more difficult than necessary. If, therefore, the document is compiled with the following criteria in mind, the compiler will be well on the way to facilitating the reviewer's task.

Very often the reviewer will wish to make notes as he reads the document so, using a large font with line spacing at 1.5 or 2 and large margins will make life easier in this respect.

Many claims are drafted in such a way that it is necessary to wade through a great deal of other paperwork to understand the case. Even if such documents are included within the claim submission, it is important that the claim narrative may be read and understood without making constant reference to such documents. A reviewer who is obliged to constantly refer to other documents to make sense of the claim is unlikely to be predisposed to the claimant. It is therefore important to resist the temptation to rely on such documents to 'tell the story' and ensure that the essential information from the documents is reproduced in the narrative. For example, rather than make a bland statement that letter reference 'abc' dated 'xyz' explains the circumstances of the delay, it is preferable to quote from the letter or paraphrase its contents within the narrative. A copy of the letter should, of course, be included in the appendices, and the narrative should include a cross-reference to its location within the appendices, so that the reviewer is able to easily locate it for verification.

When the narrative contains references to other documents or exhibits included as substantiation, it is annoying for the reviewer to have to spend time trying to locate the substantiating document or exhibits, to verify the references. Such documents should therefore be clearly arranged and labelled. If the submission is large enough to warrant presentation in more than one volume, the narrative should be contained in one volume and the exhibits in a separate volume or volumes so that the reviewer can have both the narrative and the file(s) containing the exhibits open on his desk at the same time. This will enable him to easily refer to the exhibits while reading the narrative without constantly turning pages and losing his place within a single volume.

Making the Submission or Review a Stand-Alone Document

If the reviewer has to search through his own records and files for documents referred to in the narrative in order to understand the claim or to substantiate statements made by the claimant, he will certainly not be well disposed towards the party who has caused him such unnecessary work. In the case of a claim determination, the reviewer may very well take the view that a failure by the claimant to properly substantiate the claim means that the claim has not been proven and therefore the reviewer may simply reject the claim as submitted.

It could be that a reviewer is not familiar with the project or does not have ready access to the project records. In such a case, he will be unable to complete his task without requesting additional particulars from the parties. This is another annoyance and, of course, will serve to delay the conclusion of the response or determination. For this reason it is necessary to include *absolutely everything* that the reviewer will need to refer to within the submission documents. In other words, ensure that the submission is a completely stand-alone document. Submissions should include:

- copies of correspondence referred to in the narrative
- copies of other project records referred to in the narrative or used in calculations
- copies of drawings used as substantiation or as a basis of calculations
- relevant extracts from the conditions of contract
- programmes used in substantiation of the claim
- calculations used to evaluate quantum
- copies of invoices in substantiation of prices.

Obviously, in some cases it will be impractical to include large or unwieldy documents within the submission and in such cases it is acceptable that the reviewer should refer to such documents separately.

Chapter 4

Such omissions should, however, be the exception rather than the rule. Examples of these are as follows:

- The original submittal, in the case where the document under preparation is a review or determination
- Previous interim submissions on the same subject
- Large or numerous drawings; it is advisable, however, to include copies of the title block showing the name, drawing number, date and revision and, if appropriate, the section of the drawing that is referred to in the submission
- The complete conditions of contract

Superfluous and Irrelevant Information

While it is essential that the claim or response should comprise a stand-alone document as discussed above, many claims actually contain *too much* information. I get the distinct impression that the compilers of documents that fall into this category take the view that if they include as much information as possible, the reviewer will happily sort through it all to find out what is relevant. This of course does not comply with the requirements to make the reviewer's job as easy and as pleasant as possible.

Very often, the person who is tasked with compiling the claim will have been presented with, or will have gathered, a vast amount of information, some of which is relevant and some of which is not. One of a good claim practitioner's skills is to be able to review such information and to decide what is relevant and is to be used and what is not relevant and should be discarded. It is very tempting to attempt to put all available information within a submission, but any discussion of matters that do not have a direct bearing on the issue in question should be avoided.

The inclusion of irrelevant information serves no purpose at all and will do nothing to bring about a clear understanding of the matters being discussed. In fact, the reverse is true and a reviewer could very well suffer from information overload. Narratives should therefore be as concise as possible while still properly illustrating and explaining the points being made. Exhibits and appendices used to support the document should also be kept to a minimum while still being sufficient to comprise a stand-alone document.

Always remember that quantity is never a substitute for quality.

Do not Assume that the Reviewer has Prior Knowledge of the Project or Circumstances

In most cases a claim submission or review will be dealt with, at least in the first instance, by people who have everyday knowledge of the project and of the circumstances surrounding the claim and in many cases the issue will probably remain at this level. It must be borne in mind, however, that this will not always be the case. In a situation that includes large or complicated claims, it could be that the project personnel may refer the issue to other more experienced or qualified persons who have little or no familiarity with the project. If the issue progresses to a dispute, then it is even more likely that the people called on to become involved in the issue will have little or no prior knowledge of the project or the circumstances and events leading to the claim. For this reason, in addition to the claim or review document being of a stand-alone nature in terms of the contents and supporting documentation, it is also important to compile the narrative and prepare the submission so as to include enough information and present it in such a way so that someone with absolutely no prior knowledge of the issues may fully understand it. It may seem unnecessary to include, for example, a project description or the names of the parties in the narrative, but firstly this has to be done only once for it to be cut and pasted into subsequent documents and secondly this information will be important, if not essential, to an arbitrator or suchlike. A side benefit, in presenting claim documents in such a way, is that the claimant is in effect intimating that he is confident in his case and if the reviewer does not agree, then the claimant already has a comprehensive claim submission ready and waiting to take to the next level.

Chapter 4

The Importance of Leading the Reviewer to a Logical Conclusion

Every good story has a beginning, a middle and an end, and the same is true of a claim or review narrative. The purpose of the narrative is to firstly set the scene by explaining the background of the project and the general circumstances; this could be regarded as the beginning. The middle of the story is where the action takes place and in the case of a claim narrative, this is where we explain the events that have occurred and discuss cause, effect and entitlement. If the narrative author has done a good job, this will in turn lead to an ending which will be a logical conclusion in which the entitlement and quantum are summarised. As in all good novels, if this premise is adhered to, the heroes will hopefully live happily ever after.

In the case of a claim review, the reviewer may be somewhat constrained in achieving this if he decides to follow the same order of presentation for a claim that is not presented in a very logical manner. The reviewer needs to decide whether his purposes will be served better by following the order of the claimant's presentation or, if he feels this would be more logical, by responding in his own way.

Explanations, Summaries and Conclusions

The purpose of a claim or response is to convince the reviewer that the claim or determination is accurate and just, and the majority of the work required to achieve this situation is by way of the narrative.

Many narrative writers do a good job of presenting the facts, but fail to lead the reviewer to the conclusion that is desired. This is very dangerous, because if the claim or response does not clearly tell the reviewer what the outcome of the claim or response should be, the reviewer may very well reach a totally different conclusion.

One way in which this may be achieved is by including explanations, summaries and conclusions at regular intervals within the narrative. Simply put, we should state the facts, explain what the facts mean, summarise what has been said and then tell the reader what conclusion must be drawn in terms of the claim or response.

Use of the Narrative to Explain other Documents

When reviewing a claim it is often very obvious that the document has been prepared by more than one person. In such instances the final result is usually somewhat disjointed and there is often little interrelationship between the separate parts of the submission. Possibly, one person has been responsible for the narrative, someone else has prepared programmes to demonstrate the effect of delays and a third person has calculated the additional payment claimed. In such cases, the three separate documents may be perfectly well prepared and presented on an individual basis, but there may be inconsistencies between the individual documents. Additionally, it is quite common for there to be little in the way of explanation of the logic behind such documents, or the basis of their preparation. Possibly, to another planner, the programme may be understandable and another quantity surveyor may be able to follow the calculations, but to a non-expert in these particular fields, this may prove to be somewhat of a challenge. Additionally, names, references and descriptions may be inconsistent across the various documents. For example, the narrative section may refer to 'the Quantum Calculations' and yet the spreadsheets that calculate the quantum may have a heading 'Additional Costs'. If they are the same thing, they should have the same name. Bearing in mind that we are trying to make life easy for the reviewer, it is also essential to provide detailed explanations of calculations and the like within the narrative in such a way that a non-expert may understand the logic.

If, for example, a baseline programme is used to demonstrate the effect of delays by production of an impacted baseline programme, then the narrative should explain in step-by-step detail each change made to the activities and why this has been done. Revised dates and durations should, if necessary, be supported by calculations.

Spreadsheets included within the submission to demonstrate calculations and evaluations should be supported by detailed explanations and, if necessary, calculations should be reproduced in the narrative. Column headings in spreadsheets should always provide essential information as to what the underlying information represents. In short, the reviewer should never have to resort to a calculator in order to make sense of the calculations contained therein.

Substantiation by the use of Exhibits and Additional Documents

It is never sufficient to make a statement without substantiation of its veracity. Equally, if the reviewer of a document is doing his job correctly, he will need to satisfy himself as to the correctness of the statements made. Let's take two examples of how a claim for a simple variation may be presented.

Example No. 1

On 2 February 2010, the Engineer issued revised drawings to show an additional staircase between the Ground Floor and Basement Level 1.

The work detailed on the revised drawings constitutes a variation which has resulted in additional concrete works, masonry, metalwork and decoration.

The additional price to which the Contractor is entitled is £5,321.

A review of this immediately raises the following questions:

- How were the revised drawings issued?
- Which drawings were revised and what are the revision numbers?
- Why does this instruction constitute a variation?
- Why does this result in an entitlement to additional payment?
- How has the claimed sum been calculated?

A reviewer would therefore have difficulty in reaching a determination in the absence of such vital information. The following example shows a more satisfactory approach:

Chapter 4

Example No. 2

On 2 February 2010[1], the Engineer issued revised drawings[2,3] to show an additional staircase between the Ground Floor and Basement Level 1.

Sub-Clause 3.3 [*Instructions of the Engineer*][4] provides that the Engineer may issue modified drawings and Sub-Clause 13.1 [*Right to Vary*][5] provides that the Engineer may instruct a variation for additional work to be evaluated under Clause 12 [*Measurement and Evaluation*][6].

Examination of the Contract Drawings[7,8] shows that the staircase detailed on the revised drawings is additional to the Contract and thus constitutes a variation. The provision of the staircase has resulted in additional concrete works, masonry, metalwork and decoration, the quantities of which are included herewith under Appendix B.

The evaluation of the variation is included herewith under Appendix A and demonstrates that the additional price to which the Contractor is entitled is £5,321.

---

[1] Exhibit 1 – Drawing Issue No. D 023, dated 2 February 2010
[2] Exhibit 2 – Drawing No. A-GF-1001 Rev A
[3] Exhibit 3 – Drawing No. A-B1–1001 Rev A
[4] Exhibit 4 – Extract from the Conditions of Contract; Sub-Clause 3.3
[5] Exhibit 5 – Extract from the Conditions of Contract; Sub-Clause 13.1
[6] Exhibit 6 – Extract from the Conditions of Contract; Clause 12
[7] Exhibit 2 – Drawing No. A-GF-1001 Rev 0
[8] Exhibit 3 – Drawing No. A-B1–1001 Rev 0

The narrative contained in this example not only answers all the questions in the reviewer's mind, but also provides substantiation by reference to the various documents.

For ease of use, the referenced documents in the second example should be included in the submission document as either exhibits or appendices and should be cross-referenced in the narrative by way of footnotes as shown.

Compilation of the Document

In this chapter we have discussed the following:

• Professional quality document by the use of quality presentation materials and a good writing style

- Making the submission user-friendly
- Making the submission a stand-alone document by including every-thing to which the reviewer will need to refer
- Writing the narrative on the assumption that the reviewer has no pre-vious knowledge of the project
- Leading the reviewer to a logical conclusion
- The use of narrative to provide explanations of other documents included in the submission
- The use of exhibits and additional documents as substantiation

In order to gain the maximum advantage from the above, it is neces-sary that the content of the submission document is compiled in a logi-cal manner and in a way that is user-friendly. A typical way in which to order the document for a claim for an extension of time and additional payment for prolongation would be as follows:

1. Volume 1
   a. Front cover
   b. Contents
   c. Narrative
2. Volume 2
   a. Front cover
   b. Contents
   c. Appendix A – Exhibits, i.e. documents used as substantiation
   d. Appendix B – Baseline Programme
   e. Appendix C – Entitlement Programme
   f. Appendix D – Calculation of Additional Costs
   g. Appendix E – Supporting Information for the Cost Calculations

It should be noted that, in this example, the submission is compiled in two volumes with the narrative contained in Volume 1. This is to ena-ble the reviewer to refer easily to the supporting documents contained in Volume 2 while reading the narrative. Each section or appendix should have dividers with clearly labelled tabs; if necessary, the sec-tions or appendices should have sub-dividers to assist in the location of documents. For example, if Appendix A contains exhibits referenced 1 to 20, then each individual exhibit should be located behind a sub-divider with a tab and an appropriate label.

As may be imagined from the foregoing, the final submission will be a substantial document rather than the few pieces of dog-eared paper referred to earlier. This very fact brings about another advantage to the claimant, which is informally known as the 'thud factor'. The 'thud factor' is based upon the premise that when a document lands on someone's desk with a loud thud, then that person is likely to conclude that the party who has prepared the document, firstly knows what they are doing and secondly is taking the matter seriously.

Summary of the Principles Covered in this Chapter

- Make the reviewer's job as easy and as pleasant as possible.
- Ensure that the submission document is well presented.
- Ensure that the document is user-friendly.
- Ensure that the submission is a stand-alone document.
- Avoid superfluous and irrelevant information.
- Assume that the reviewer has no prior knowledge of the project.
- Use the narrative to lead the reviewer to a logical conclusion.
- Include explanations, summaries and conclusions to good effect within narratives.
- Use the narrative to explain other documents attached as substantiation or in support of the narrative.
- Ensure that wording, titles and the like included in supporting documents are consistent with the narrative.
- Ensure that the logic contained in supporting calculations, programmes and so on is explained clearly.
- Ensure that statements made are substantiated by reference to the project records or other documents, and include copies of such documents as substantiation.
- Take care with prose, grammar and punctuation and ensure that the narrative can be easily read and properly understood.
- Avoid the use of acronyms and abbreviations.
- Keep the writing style simple and direct.
- Avoid the use of legalese and unnecessarily complicated language.
- Ensure that references to the parties within the narrative are unambiguous.
- Identify quotations correctly and consistently.
- When possible, use the actual wording of clauses rather than paraphrasing their meanings.
- Ensure that the submission document is well ordered and indexed to enable a reviewer to quickly find documents.
- Present reference material and documents used as substantiation in a separate volume to the narrative.
- Ensure that an in-house review is carried out before finalisation of the document.

Chapter 4

Chapter 5

Essential Elements of a Successful Claim

Introduction

The previous chapter dealt with *how* a claim submission or response document should be presented, and in this chapter we will get to the more serious matter of *what* should be presented. While it is true that a badly presented but robust case will be weakened by poor presentation, it is also accurate to say that the most attractive, well-ordered and professionally written document will fail if it does not contain the necessary substance.

The object of a claim is to demonstrate that on the balance of probability the claimant is entitled to compensation and also to demonstrate and substantiate the amount of such compensation. In the case of a construction claim, the compensation would more than likely consist of additional time, additional payment or both. The claim must be demonstrated, substantiated and justified so as to achieve the desired result. The elements that are *absolutely essential* to include in a claim or determination are as follows:

1. Cause
2. Effect
3. Entitlement
4. Substantiation

It may help to remember these essentials by use of the acronym CEES.

It should definitely be borne in mind that many worthwhile claims have come to nothing because these essentials have not been given due recognition by the claimant.

We will examine CEES in detail in the remainder of this chapter and also demonstrate, by way of a practical example, how these elements may be dealt with in a typical claim for an extension of time. The example we will use is based on the following scenario:

Construction Claims & Responses: Effective Writing & Presentation, Second Edition. Andy Hewitt.
© 2016 John Wiley & Sons, Ltd. Published 2016 by John Wiley & Sons, Ltd.

1. The Employer has entered into a contract with BBO Construction to construct 85 two-storey, detached dwellings. The contract is the FIDIC *Contract for Building and Engineering Works designed by the Employer 1999 Edition*.
2. The Employer has also engaged a separate contractor to construct the infrastructure for the development, which includes mains drainage, electricity distribution, telephone ducting, roads, pavements and public landscaping.
3. The two contracts are running concurrently and the Engineer is the same party for both contracts.
4. On 1 February 2010, the infrastructure contractor excavated a trench for a road crossing across the road leading to four of the dwellings and prevented the Contractor from accessing these dwellings.
5. The infrastructure contractor completed the road crossing and backfilled the trench on 9 February 2010.
6. The Contractor considers that the lack of access caused by the road crossing delayed his work to the four dwellings and that this delay had a direct effect on the completion of the project and, consequently, he is entitled to an extension of time for this delay.

Cause

Very simply put, the Cause is the event that has given rise to the claim. Typically, this could be:

- late or restricted access to the site
- the issue of an instruction to carry out additional work
- the issue of a revised drawing
- late issue of instructions or information
- the issue of an instruction to suspend the works
- the issue of an instruction to accelerate the works
- exceptionally adverse climatic conditions
- delays caused by local or statutory authorities
- changes in government legislation
- force majeure
- delay caused by the Employer or other parties engaged by the Employer
- an act of prevention by the Employer, his agents or contractors.

The Cause is generally a statement of fact and is usually fairly easily to establish by way of the project records. In our example, the Cause is the fact that the infrastructure contractor excavated the trench for the road crossing. In a typical claim narrative, the Cause could be explained as follows:

The Cause

1. On 1 February 2010 the Employer's infrastructure contractor excavated a trench across the road leading to house numbers 36, 38, 40 and 42. These houses may only be accessed by way of the road that was affected by the infrastructure contractor's works.
2. The infrastructure contractor's work was completed and access to the four affected houses was re-established on 9 February 2010.

Effect

For a claim to succeed it will be necessary to demonstrate that the Effect on which the claimed compensation is based was, in fact caused by the event, by linking the Cause with the Effect. The Effect of the event is usually a little more complicated to establish and to link directly to the Cause, because this is often a subjective matter that requires to be both demonstrated and substantiated. The following are examples of what should be considered when examining the Effect of an event:

Time considerations

1. Will the event cause delay because of additional or changed work?
2. Will the event cause delay because of its timing?
3. What effect will the delay have on the programme?
4. Will the delay affect the completion date?
5. Will the event require acceleration measures
6. Does the event fall under the contractual provisions that allow an extension of time to be claimed?

Financial considerations

1. Will a delay cause the Contractor to incur additional costs due to prolongation of the works or a part of the works?
2. Will compliance with an instruction require rework, or result in abortive work having been carried out?
3. How will the payment for additional work and any abortive work or rework be claimed?
4. How will the adjustment to the contract price for additional work be made? Is the contract remeasurable or a lump sum?
5. If there are requirements to mitigate delays, would the mitigation measures give rise to entitlement to additional payment and how would these be paid?

Chapter 5

6. Will the event result in idle or down time of resources and how would the Contractor be compensated for incurred costs?

7. Will any acceleration measures require increased resources and, if so, what additional costs will be incurred in mobilising and maintaining such resources and how would the Contractor be compensated?

8. Will increased production resources for acceleration also require additional management, supervision, administration, accommodation, transport resources and so on to support them and, if so, what are the costs and how will they be paid?

9. Will acceleration measures affect the costs of production and, if so, what will be the effect on the costs?

10. In the case of a suspension, is there a requirement to protect the works during the period of suspension and, if so, what additional costs will be incurred?

11. Will the event require the demobilisation and remobilisation of resources and, if so, what additional time and costs will be incurred in doing so and how would these be compensated?

12. Does the event fall under the contractual provisions that allow for additional payment to be claimed?

The above forms a useful checklist, which, if considered against any event, will provide guidance to what the effect of the event may be.

You will hopefully notice that all the above considerations have been written in the present tense. This is because the time to start considering the effect is immediately that there is knowledge of the event. This is the point in time that the contract administration procedures should kick in to ensure that notices are sent, records are kept and preparations are made to submit the particulars of the claim if subsequent investigations show that an actual delay to the Time for Completion has occurred.

Returning now to our example, the following narrative would demonstrate the Effect of the Cause discussed earlier in this chapter:

Effect

1. The road crossing restricted the Contractor's access to the four dwellings by preventing vehicles from reaching the houses to deliver the construction materials necessary for progress to be maintained.

2. The infrastructure contractor commenced excavation of the road crossing on 1 February 2010, which effectively restricted access from this day. On this date, the progress of the individual dwellings was as follows:

   a. House No. 36 – reinforcement to the raft foundation was in progress and due to be completed on 2 February 2010

   b. House No. 38 – ready for concrete to be poured to the raft foundation

   c. House No. 40 – blockwork to ground-floor external walls and partitions was in progress and due to be completed on 3 February 2010

   d. House No. 42 – blockwork completed to ground floor. First-floor precast concrete flooring beams due for delivery and placement on 1 February 2010

3. Progress to the affected dwellings was dependent on the delivery of ready-mixed concrete, concrete blocks, cement, sand, precast concrete flooring beams and other materials to the working areas. No other access route for such vehicles was available due to other activities and completed construction works in this area of the site. The effect of the excavation of the road crossing by the infrastructure contractor was to prevent deliveries from being made between the dates of 1 and 9 February 2010 and, consequently, to suspend the construction activities until 10 February 2010, the day after the access was reinstated.

4. This period of suspension had the effect of delaying the Time for Completion of the Works, and this has been demonstrated by impacting the event on the individual activities on the current baseline programme (included herein under Appendix A), in order to produce an impacted baseline programme which is included herein under Appendix B.

5. The effect on each dwelling is shown as follows:

   a. House No. 36

      i. Progress on 1 February 2010: reinforcement to the raft foundation in progress and due to be completed on 2 February 2010.

      ii. The concrete gang can only complete one raft per day and House No. 38 was programmed to start prior to House No. 36. Thus, concreting to House No. 38 took place on 10 February 2010 when the access was reinstated. The gang followed on with House No. 36 on 11 February 2010. Thus, the effect was to delay concreting of the raft foundation from the planned date of 3 February to 11 February 2010, a delay of 8 calendar days.

      iii. The effect on the overall programme has been demonstrated by impacting the baseline programme as follows:

       1. Activity: concrete to raft foundation

       2. Activity start date deferred by 8 calendar days

  b. House No. 38

    i. Progress on 1 February 2010: ready for concrete to be poured to the raft foundation on that day.

    ii. The effect was to delay concreting of the raft foundation from the planned date of 1 February to 10 February 2010, a delay of 9 calendar days.

    iii. The effect on the overall programme has been demonstrated by impacting the baseline programme as follows:

       1. Activity: concrete to raft foundation

       2. Activity start date deferred by 9 calendar days

  c. House No. 40

    i. Progress on 1 February 2010: blockwork to ground-floor external walls and partitions was in progress and due to be completed on 3 February 2010.

    ii. The effect was to suspend progress on the ground-floor blockwork from 1 February to 10 February 2010 and thus prevent completion until 12 February 2010, a delay of 9 calendar days.

    iii. The effect on the overall programme has been demonstrated by impacting the baseline programme as follows:

       1. Activity: blockwork to ground floor

       2. Activity duration increased by 9 calendar days

  d. House No. 42

    i. Progress on 1 February 2010: blockwork completed to ground floor. First-floor precast concrete flooring beams due for delivery and placement on 1 February 2010.

    ii. The effect was to delay the installation of the first-floor precast concrete flooring from 1 February to 10 February 2010, a delay of 9 calendar days.

    iii. The effect on the overall programme has been demonstrated by impacting the baseline programme as follows:

       1. Activity: PCC flooring

       2. Activity start date deferred by 9 calendar days

6. Reference to the baseline programme included herein under Appendix A shows that this cluster of four dwellings was the last to be started and thus the last to be completed. The effect of this delay event therefore had a direct effect on the Time for Completion of the project. The impacted baseline programme included herein under Appendix B demonstrates that the effect on the individual activities of the affected dwellings has had the overall effect of delaying the Time for Completion by 9 days, i.e. until 6 August 2010.

In the above narrative we have taken the basic cause, i.e. the fact that the infrastructure contractor excavated a road crossing, and we have developed the facts of the matter to demonstrate the effect, not only on the Contractor's working arrangements by discussing which activities were affected and in what way, but also on the Contractor's programme. Finally, we have demonstrated the extension of time to which the Contractor is entitled by impacting the delays into the programme. Returning to the analogy that every good story has a beginning, a middle and an end, this would be the middle of the story where the action takes place.

Entitlement

A demonstration of the cause and effect of an event will not automatically contain entitlement to an extension of time and/or additional payment. The claim will flow either from a remedy contained in the contract conditions, or from a breach of the contract giving rise to common law damages and could possibly fall under both categories. It is of vital importance to set out precisely on what contractual basis the claim is made.

A substantial part of any contract is the allocation of risk between the parties and it is therefore necessary to demonstrate that the event on which the claim is based is something for which the contract, or the law to which the contract is subject, provides entitlement to the claimant. In the case of our example, it could very well be that the Contract contains obligations for the Contractor to coordinate activities with other contractors employed by the Employer and allocates the associated risks to the Contractor. In such a case, it would be more difficult to demonstrate the Contractor's entitlement to an extension of time for the case in question. It is, therefore, imperative to state precisely on what contractual basis the claim is founded.

The first place to check for entitlement is the Contract. If we examine FIDIC in connection with our example, we will see that the event upon which our example is based is covered under Sub-Clause 8.4 [*Extension of Time for Completion*], which contains specific provisions dealing with delays caused by the Employer's other contractors as follows:

The Contractor shall be entitled subject to Sub-Clause 20.1 [*Contractor's Claims*] to an extension of the Time for Completion if and to the extent that completion for the purposes of Sub-Clause 10.1 [*Taking Over of the Works and Sections*] is or will be delayed by any of the following causes:

(a) a Variation (unless an adjustment to the Time for Completion has been agreed under Sub-Clause 13.3 [*Variation Procedure*]) or other substantial change in the quantity of an item of work included in the Contract,

(b) a cause of delay giving an entitlement to extension of time under a Sub-Clause of these Conditions,

(c) exceptionally adverse climatic conditions,

(d) Unforeseeable shortages in the availability of personnel or Goods caused by epidemic or governmental actions, or

(e) any delay, impediment or prevention caused by or attributable to the Employer, the Employer's Personnel, or the Employer's other contractors on the Site.

If the Contractor considers himself to be entitled to an extension of the Time for Completion, the Contractor shall give notice to the Engineer in accordance with Sub-Clause 20.1 [*Contractor's Claims*]. When determining each extension of time under Sub-Clause 20.1, the Engineer shall review previous determinations and may increase, but shall not decrease, the total extension of time.

Chapter 5

Examination of the above reveals that sub-paragraph (e) provides the Contractor with entitlement to an extension of time in the case of delays by the Employer's other contractors on the Site.

Although our example is fairly straightforward when reviewed against the FIDIC conditions of contract, it is sometimes the case that entitlement is not so clear cut. In such cases, persuasive arguments including expert opinion, and case law may have to be brought into play in order to sway the balance. On the other hand, there are some events that would fall under more than one clause that would give entitlement. In such cases, it would be better to examine and include all such provisions within the claim submission. The latter may be a 'belt and braces' approach, but the small amount of additional time taken to strengthen the case can be worthwhile, especially if the reviewer subsequently finds flaws in one of the reasons put forward to establish entitlement.

Bearing in mind the two principles that it is incumbent on the claimant to prove the merits of the case and that we have to do everything we can to make the reviewer's job as easy as possible, the claim submission must contain a clear demonstration of the claimant's entitlement by reference to the Contract. Returning to our example, the following narrative would demonstrate the claimant's entitlement for the case in question:

The Contractor's Entitlement

1. Extension to the Time for Completion
   a. The Contractor's entitlement to an extension to the Time for Completion is contained within the provisions of Sub-Clause 8.4 [*Extension of Time for Completion*] which, under sub-paragraph (e), provides that an extension of time shall be given for delay caused by the Employer's other contractors on the Site.
   b. The event of the road closure by the infrastructure contractor, as described herein, clearly falls under this provision and, consequently, the Contract provides that the Contractor shall be entitled to an extension of the Time for Completion if and to the extent that completion is or will be delayed. The claim submitted herein contains the Contractor's request for an extension of Time for Completion for the nine days of delay demonstrated.

2. Additional Payment
   a. Sub-Clause 8.4 [*Extension of Time for Completion*] also contains a reference to Sub-Clause 20.1 [*Contractor's Claims*] which provides that the Contractor may also claim additional payment for circumstances which cause an extension for the Time for Completion to be awarded.
   b. Due to the circumstances presented herein, the Contractor was obliged to remain on site for a period greater than was originally intended and thereby incurred additional costs in maintaining his site establishment, providing finance for the works and maintaining head-office overheads during the extended period. He was also prevented from earning a contribution from other projects through having his resources tied up for the extended period.
   c. The principles of recovery where one party to a contract has defaulted are well established in law. Essentially, the aggrieved party is entitled by an award of money to be put back in the position in which it would have been had the contract been performed as originally envisaged.
   d. Sub-Clause 20.1 [*Contractor's Claims*] also provides that the Engineer is obliged to respond with approval, or with disapproval and detailed comments to the submitted claim and subsequently that payment certificates shall include such amounts for any claim that are due under the relevant provision of the Contract.
   e. It is therefore the Contractor's further claim that due to the circumstances entitling him to an extension of the Time for

Completion of the Works, the Contractor is also entitled pursuant to both the Contract and to common law to additional payment to recompense him for the costs incurred as a result of the additional time he has been obliged to remain on site. The Contractor's claim in this respect will be submitted by way of a separate claim.

3. Liquidated Damages or Penalties
   a. The Employer has an entitlement to deduct penalties for late completion which is contained under Sub-Clause 8.7 [*Delay Damages*]. This sub-clause, however, contains a reference to Sub-Clause 8.2 [*Time for Completion*] which provides that the Contractor shall complete the whole of the Works and each Section (if any) within the Time for Completion for the Works.
   b. As has been examined earlier herein, the Time for Completion may, however, be extended under the provisions of Sub-Clause 8.4 [*Extension of Time for Completion*] if and to the extent that completion is or will be delayed.
   c. Thus, the entitlement of the Employer to the payment of delay damages is negated for any circumstances that entitle the Contractor to an extension of the Time for Completion. As is demonstrated herein, the Contractor is entitled to such an extension of time and therefore the Employer is not entitled to the payment of delay damages by the Contractor.

4. Conditions Precedent to Entitlement
   a. Sub-Clause 20.1 [*Contractor's Claims*] provides that the Contractor is obliged to give notice to the Engineer describing the event or circumstance giving rise to the claim and that the notice shall be given as soon as practicable and not later than 28 days after the Contractor became aware, or should have become aware, of the event or circumstance.
   b. If the Contractor fails to give notice of a claim within 28 days, the Contractor is not entitled to additional time or payment, and the Employer is discharged from all liability in connection with the claim.
   c. This sub-clause also provides that, within 42 days after the Contractor became aware of the event or circumstance giving rise to the claim, or within such other period as may be proposed by the Contractor and approved by the Engineer, the Contractor shall send to the Engineer a fully detailed claim which includes full supporting particulars of the basis of the claim and of the extension of time and/or additional payment claimed.
   d. Thus, the requirement to give notice of entitlement to an extension of time and to describe the event giving rise to the

claim within 28 days of the event is a condition precedent to the Contractor's entitlement. The Contractor is also obliged to submit a fully detailed claim of the extension of time claimed within 42 days of the event.

e. The Contractor submitted a notice of claim on 15 February 2010 which is within the 28-day period prescribed in the Contract. The submission contained herein comprises the detailed claim and supporting particulars, thereby satisfying the provisions of this sub-clause.

f. The Contractor has therefore complied with the conditions of this sub-clause and is consequently entitled to an extension of the Time for Completion until 6 August 2010.

5. Conclusion

   a. The following is a summary of the Contractor's entitlement as discussed in this section:

     i. The Contractor is entitled under the Contract to an extension to the Time for Completion for delay, impediment or prevention caused by or attributable to the Employer's other contractors on the Site.

     ii. Due to the circumstances entitling the Contractor to an extension of the Time for Completion, the Contractor is also entitled, pursuant to both the Contract and the law, to additional payment for the costs incurred as a result of the additional time he has been obliged to remain on site. The Contractor's claim in this respect will be submitted by way of a separate claim.

     iii. The Employer is not entitled to the payment of delay damages by the Contractor.

     iv. The Contractor has complied with the conditions precedent to entitlement.

Substantiation

The last essential element of the claim is substantiation or, in other words, proving to a reasonable level that all statements made, points relied on, calculations submitted and the like are correct. Imagine a prosecution lawyer during a murder trial standing up in court and making a simple statement to the effect that the defendant is guilty of the crime of which he is accused because he was at the location of the crime when it took place and that he didn't like the victim very much. Is it likely that the jury would take this at face value and convict the accused on the basis of such a statement, or is it more likely that they would need some sort of proof of the accusations? The answer is fairly obvious and the lawyer will consequently take his time to substantiate each and every one of his assertions by reference to the evidence that he has gathered to enable him to prove his case.

In order to prove or substantiate the events and circumstances on which the claim is based, it is essential that a claim submission contains similar levels of evidence as the lawyer would use in his trial. This also has to be done by the use of evidence, which, in most cases, may be obtained from the project records.

Let us now examine how we can enhance the claim narrative in the same way that the trial lawyer would submit evidence to the jury, in order to attempt to prove his case. If we re-examine sections of the narrative developed earlier in this chapter, we can see that thus far the narrative is merely a collection of statements as follows:

The Cause

1. On 1 February 2010 the Employer's infrastructure contractor excavated a trench across the road leading to house numbers 36, 38, 40 and 42. These houses may only be accessed by way of the road that was affected by the infrastructure contractor's works.
2. The infrastructure contractor's work was completed and access to the four affected houses was re-established on 9 February 2010.

Let's now see how we can substantiate the above by including evidence into the narrative as follows:

The Cause

1. On 1 February 2010, the Contractor wrote to the Engineer to advise that the Employer's infrastructure contractor had excavated a trench across the road leading to house numbers 36, 38, 40 and 42.[1] These houses may only be accessed by way of the road that was affected by the infrastructure contractor's works. Photographs taken on the same day are included herein under

---

[1] Exhibit 1 – BBO Construction letter reference BBOC/P9921/L0347, dated 01/02/10

Appendix A and show the extent of the infrastructure contractor's work and the restricted access.

2. It was recorded in the site meeting held on 10 February 2010 that the infrastructure contractor's work was completed and access to the four affected houses was re-established on 9 February 2010.[2]

---

[2] Exhibit 2 – Site Meeting Minutes, dated 10/02/10

The above example introduces evidence to substantiate the statements by reference to the project records, in this case: correspondence, photographs and meeting minutes. The exhibits referred to as evidence or substantiation via the footnotes should be appended to the claim submission and clearly indexed for ease of reference. The use of footnotes to reference the exhibits rather than quoting the document reference numbers within the text allows the narrative to flow. This makes the task of reading more pleasant and provides for an easier understanding of the text. It is sensible to allocate individual exhibit numbers to each document because it is often necessary to refer to a document on several occasions within the narrative and by adopting this method the document only needs to be included once in the appendices.

The usual project records of correspondence, minutes, site reports and so on can, and should, be supplemented by additional records in order to substantiate the events and the subsequent effect. The following is a section from the original example narrative that deals with the effect:

The Effect

1. The road crossing restricted the Contractor's access to the four dwellings by preventing vehicles from reaching the houses to deliver the construction materials necessary for progress to be maintained.

2. The infrastructure contractor commenced excavation of the road crossing on 1 February 2010, which effectively restricted access from this day. On this date, the progress of the individual dwellings was as follows:
   a. House No. 36 – reinforcement to the raft foundation was in progress and due to be completed on 2 February 2010.

b. House No. 38 – ready for concrete to be poured to the raft foundation.

c. House No. 40 – blockwork to ground-floor external walls and partitions was in progress and due to be completed on 3 February 2010.

d. House No. 42 – blockwork completed to ground floor. First-floor precast concrete flooring beams due for delivery and placement on 1 February 2010.

3. Progress to the affected dwellings was dependent on the delivery of ready-mixed concrete, concrete blocks, cement, sand and precast concrete flooring beams and other materials to the working areas. No other access route for such vehicles was available due to other activities and completed construction works in this area of the site. The effect of the excavation of the road crossing by the infrastructure contractor was to prevent deliveries from being made between the dates of 1 and 9 February 2010 and, consequently, to suspend the construction activities until 10 February 2010, the day after the access was reinstated.

Now let us look at this section again to see how the use of drawings and photographs, made and recorded, specifically with respect to the claim event, may be used to provide substantiation.

The Effect

1. Appendix B contains a site plan, which has been marked up to show the location of the road crossing. The plan also shows that alternative access to the dwellings in question was not possible, due to the location of existing boundary walls and other completed works. The photographs contained in Appendix A also demonstrate that the works shown on the site plan had already been constructed at the time in question. Thus, the road crossing restricted the Contractor's access to the four dwellings by preventing vehicles from reaching the houses to deliver the construction materials necessary for progress to be maintained.

2. The infrastructure contractor commenced excavation of the road crossing on 1 February 2010, which effectively restricted access

from this day. The Daily Site Report of 1 February 2010[3] and the photographs in Appendix A record that the progress of the individual dwellings was as follows:

   a. House No. 36 – reinforcement to the raft foundation was in progress and due to be completed on 2 February 2010.
   b. House No. 38 – ready for concrete to be poured to the raft foundation.
   c. House No. 40 – blockwork to ground-floor external walls and partitions was in progress and due to be completed on 3 February 2010.
   d. House No. 42 – blockwork completed to ground floor. First-floor precast concrete flooring beams due for delivery and placement on 1 February 2010.

3. Progress to the affected dwellings was dependent on the delivery of ready-mixed concrete, concrete blocks, cement, sand, precast concrete flooring beams and other materials to the working areas. No other access route for such vehicles was available due to other activities and completed construction works in this area of the site. The effect of the excavation of the road crossing by the infrastructure contractor was therefore to prevent such deliveries from being made between the dates of 1 and 9 February 2010 and effectively to suspend the construction activities until 10 February 2010; the day after the access was reinstated. The Daily Site Report of 9 February 2010[4] and the photographs included in Appendix C herein record the progress of the affected dwellings at the time when the road crossing was reinstated and the works were able to recommence as normal.

---

[3] Exhibit 3 – Daily Site Report, dated 01/02/10
[4] Exhibit 4 – Daily Site Report, dated 09/02/10

In the above example, a drawing has been produced and referred to in the narrative to demonstrate that, following the excavation of the road crossing, there was no alternative access available for the delivery of the materials necessary for progress. Photographs have also been taken and included in the claim submission in order to substantiate both the access restrictions and the progress of the affected dwellings at the time of the delay. The drawing and the photographs have been produced in addition to the normal project records to support the claim. This is a good demonstration of the necessity of having robust contract administration procedures in place, to ensure that such records are taken and maintained. In this example, had the Contractor not realised that photographs would have been useful until he started to prepare his claim, it would have obviously been too late and the opportunity would have been lost.

As discussed previously, when considering entitlement matters, the best place to look for substantiation is in the Contract. The following extract from our example does this to some extent:

The Contractor's Entitlement

1. Extension to the Time for Completion
   a. The Contractor's entitlement to an extension to the Time for Completion is contained within the provisions of Sub-Clause 8.4 [*Extension of Time for Completion*] sub-paragraph (e), which provides that an extension of time shall be given for delay caused by the Employer's other contractors on the Site.
   b. The event of the road closure by the infrastructure contractor, as described herein, clearly falls under this provision and, consequently, the Contract provides that the Contractor shall be entitled to an extension of the Time for Completion if and to the extent that completion is or will be delayed. The claim submitted herein contains the Contractor's request for an extension of Time for Completion for the nine days of delay demonstrated.

Let us now examine how the Contract may be used more effectively within the narrative, both to substantiate the statements made and to make the reviewer's job easier by removing the necessity for the reviewer having to constantly refer to his own copy of the Contract to verify the assertions.

The Contractor's Entitlement

1. Extension to the Time for Completion
   a. The Contractor's entitlement to an extension to the Time for Completion is contained within the provisions of Sub-Clause 8.4 [*Extension of Time for Completion*] which provides that:

'*The Contractor shall be entitled subject to Sub-Clause 20.1* [Contractor's Claims] *to an extension of the Time for Completion if and to the extent that completion for the purposes of Sub-Clause 10.1* [Taking Over of the Works and Sections] *is or will be delayed by any of the following causes:*

...

(e) any delay, impediment or prevention caused by or attributable to the Employer, the Employer's Personnel, or the Employer's other contractors on the Site.'

2. The event of the road closure by the infrastructure contractor, as described herein, clearly falls under the provision of '*delay, impediment or prevention caused by or attributable to the ... Employer's other contractors on the Site*'. Consequently, the Contract provides that the Contractor shall be entitled to '*an extension of the Time for Completion if and to the extent that completion ... is or will be delayed*'. The claim submitted herein contains the Contractor's request for an extension of Time for Completion for the nine days of delay demonstrated.

You will note that in the second entitlement narrative, the actual wording of the Contract has been used extensively; firstly by including a reproduction of the sections of the sub-clause that contain the entitlement, to avoid the reviewer having to refer to a separate document (the Contract) and secondly by using the exact wording in the subsequent explanations. The latter demonstrates how the provisions of the Contract relate to the actual circumstances of the event and emphasises the correctness of the statement being made. Such usage also gives confidence to the reviewer that the narrative is not attempting to demonstrate a point to the claimant's advantage by placing a different interpretation on the wording of the Contract to suit the circumstances.

The following narrative summarises the above into a complete demonstration of the essential elements of Cause, Effect, Entitlement and Substantiation as applied to the delay event used in our example:

The Cause

1. On 1 February 2010, the Contractor wrote to the Engineer to advise that the Employer's infrastructure contractor had excavated a trench across the access road leading to house numbers 36, 38, 40 and 42.[5] These houses may only be accessed by way of the road that was affected by the infrastructure contractor's works. Photographs taken on the same day are included herein under Appendix A and show the extent of the infrastructure contractor's work and the restricted access.
2. It was recorded in the site meeting held on 10 February 2010 that the infrastructure contractor's work was completed and access to the four affected houses was re-established on 9 February 2010.[6]

The Effect

1. Appendix B contains a site plan, which has been marked up to show the location of the road crossing. The plan also shows that alternative access to the dwellings in question was not possible due to the location of existing boundary walls and other completed construction works. The photographs contained in Appendix A also demonstrate that the construction works shown on the site plan had already been constructed at the time in question. Thus, the road crossing restricted the Contractor's access to the four dwellings by preventing vehicles from reaching the houses to deliver the construction materials necessary for progress to be maintained.
2. The infrastructure Contractor commenced excavation of the road crossing on 1 February 2010, which effectively restricted access from this day. The daily site report of 1 February 2010[7] and the photographs in Appendix A record that the progress of the individual dwellings was as follows:
   a. House No. 36 – reinforcement to the raft foundation was in progress and due to be completed on 2 February 2010.
   b. House No. 38 – ready for concrete to be poured to the raft foundation.
   c. House No. 40 – blockwork to ground-floor external walls and partitions was in progress and due to be completed on 3 February 2010.

---

[5] Exhibit 1 – BBO Construction letter reference BBOC/P9921/L0347, dated 01/02/10
[6] Exhibit 2 – Site Meeting Minutes, dated 10/02/10
[7] Exhibit 3 – Daily Site Report, dated 01/02/10

     d. House No. 42 – blockwork completed to ground floor. First-floor precast concrete flooring beams due for delivery and placement on 1 February 2010.

3. Progress to the affected dwellings was dependent on the delivery of ready-mixed concrete, concrete blocks, cement, sand, precast concrete flooring beams and other materials to the working areas. No other access route for such vehicles was available due to other activities and completed construction works in this area of the site. The effect of the excavation of the road crossing by the infrastructure contractor was to prevent such deliveries from being made between the dates of 1 and 9 February 2010 and effectively to suspend the construction activities until 10 February 2010, the day after the access was reinstated. The Daily Site Report of 9 February 2010[8] and the photographs included in Appendix C herein record the progress of the affected dwellings at the time when the road crossing was reinstated and the works were able to recommence as normal.

4. This period of suspension had the effect of delaying the Time for Completion of the Works and this has been demonstrated by impacting the event on the individual activities on the current base-line programme (included herein under Appendix A) in order to produce an impacted baseline programme which is included herein under Appendix B.

5. The effect on each dwelling is shown as follows:

    a. House No. 36

       i. Progress on 1 February 2010: reinforcement to the raft foundation in progress and due to be completed on 2 February 2010.

      ii. The concrete gang can only complete one raft per day and House No. 38 was programmed to start prior to House No. 36. Thus, concreting to House No. 38 took place on 10 February 2010 when the access was reinstated. The gang followed on with House No. 36 on 11 February 2010. Thus, the effect was to delay concreting of the raft foundation from the planned date of 3 February to 11 February 2010, a delay of 8 calendar days.

     iii. The effect on the overall programme has been demonstrated by impacting the baseline programme as follows:

        1. Activity: concrete to raft foundation.

        2. Activity start date deferred by 8 calendar days.

    b. House No. 38

       i. Progress on 1 February 2010: ready for concrete to be poured to the raft foundation on that day.

___

[8] Exhibit 4 – Daily Site Report, dated 09/02/10

      ii. The effect was to delay concreting of the raft foundation from the planned date of 1 February to 10 February 2010, a delay of 9 calendar days.

      iii. The effect on the overall programme has been demonstrated by impacting the baseline programme as follows:

         1. Activity: concrete to raft foundation.

         2. Activity start date deferred by 9 calendar days.

  c. House No. 40

      i. Progress on 1 February 2010: blockwork to ground-floor external walls and partitions was in progress and due to be completed on 3 February 2010.

      ii. The effect was to suspend progress on the ground-floor blockwork from 1 February to 10 February 2010 and thus prevent completion until 12 February 2010, a delay of 9 calendar days.

      iii. The effect on the overall programme has been demonstrated by impacting the baseline programme as follows:

         1. Activity: blockwork to ground floor.

         2. Activity duration increased by 9 calendar days.

  d. House No. 42

      i. Progress on 1 February 2010: blockwork completed to ground floor. First-floor precast concrete flooring beams due for delivery and placement on 1 February 2010.

      ii. The effect was to delay completion of the first-floor precast concrete flooring from 1 February to 10 February 2010, a delay of 9 calendar days.

      iii. The effect on the overall programme has been demonstrated by impacting the baseline programme as follows:

         1. Activity: PCC flooring.

         2. Activity start date deferred by 9 calendar days.

6. Reference to the baseline programme included herein under Appendix A shows that this cluster of four dwellings was the last to be started and thus the last to be completed. The effect of this delay event therefore had a direct effect on the Time for Completion of the project. The impacted baseline programme included herein under Appendix B demonstrates that the effect on the individual activities of the affected dwellings has had the overall effect of delaying the Time for Completion by 9 days, i.e. until 6 August 2010.

The Contractor's Entitlement

1. Extension to the Time for Completion

  a. The Contractor's entitlement to an extension to the Time for Completion is contained within the provisions of Sub-Clause 8.4 [*Extension of Time for Completion*] which provides that:

'*The Contractor shall be entitled subject to Sub-Clause 20.1* [Contractor's Claims] *to an extension of the Time for Completion if and to the extent that completion for the purposes of Sub-Clause 10.1* [Taking Over of the Works and Sections] *is or will be delayed by any of the following causes:*

...

(e) any delay, impediment or prevention caused by or attributable to the Employer, the Employer's Personnel, or the Employer's other contractors on the Site.'

b. The event of the road closure by the infrastructure contractor, as described herein, clearly falls under the provision of '*delay, impediment or prevention caused by or attributable to the ... Employer's other contractors on the Site*'. Consequently, the Contract provides that the Contractor shall be entitled to '*an extension of the Time for Completion if and to the extent that completion ... is or will be delayed*'. The claim submitted herein contains the Contractor's request for an extension of Time for Completion for the nine days of delay demonstrated.

2. Additional Payment

a. The above sub-clause contains a reference to Sub-Clause 20.1 [*Contractor's Claims*] which provides that:

'*If the Contractor considers himself to be entitled to any extension of the Time for Completion and/or any additional payment, under any Clause of these Conditions or otherwise in connection with the Contract, the Contractor shall give notice to the Engineer, describing the event or circumstance giving rise to the claim. The notice shall be given as soon as practicable, and not later than 28 days after the Contractor became aware, or should have become aware, of the event or circumstance.*'

b. This sub-clause provides that the Contractor may claim additional payment. Due to the circumstances causing a delay to the Time for Completion, the Contractor was obliged to remain on site for a period greater than was originally intended and thereby incurred additional costs in maintaining his site establishment, providing finance for the works and maintaining head-office overheads. He was also prevented from earning a contribution from other projects through having his resources tied up for the extended period.

c. The principles of recovery where one party to a contract has defaulted are well established in law. Essentially, the aggrieved party is entitled by an award of money to be put back in the position in which it would have been had the contract been performed as originally envisaged.

d. Sub-Clause 20.1 [*Contractor's Claims*] also provides that '*Within 42 days after receiving a claim ... the Engineer shall respond with approval, or with disapproval and detailed comments*' and subsequently that '*Each Payment Certificate shall include such amounts for any claim as have been reasonably substantiated as due under the relevant provision of the Contract*'.

e. It is therefore the Contractor's further claim that due to the circumstances entitling him to an extension of the Time for Completion of the Works, the Contractor is also entitled pursuant to both the Contract and to common law to additional payment to recompense him for the costs incurred as a result of the additional time he has been obliged to remain on site. The Contractor's claim in this respect will be submitted by way of a separate claim.

3. Liquidated Damages or Penalties

a. The Employer's entitlement to deduct penalties for late completion is contained under Sub-Clause 8.7 [*Delay Damages*] as follows:

If the Contractor fails to comply with Sub-Clause 8.2 [Time for Completion], *the Contractor shall subject to Sub-Clause 2.5* [Employer's Claims] *pay delay damages to the Employer for this default ...*

b. Sub-Clause 8.7 refers to Sub-Clause 8.2 [*Time for Completion*] which provides that:

The Contractor shall complete the whole of the Works, and each Section (if any), within the Time for Completion for the Works or Section (as the case may be) including:

(a) achieving the passing of the Tests on Completion, and

(b) completing all work which is stated in the Contract as being required for the Works or Section to be considered to be completed for the purposes of taking-over under Sub-Clause 10.1 [Taking Over of the Works and Sections].

c. As has been examined earlier herein, the Time for Completion may however, be extended under the provisions of Sub-Clause 8.4 [*Extension of Time for Completion*] as follows:

The Contractor shall be entitled subject to Sub-Clause 20.1 [Contractor's Claims] *to an extension of the Time for Completion if and to the extent that completion for the purposes of Sub-Clause 10.1* [Taking Over of the Works and Sections] *is or will be delayed.*

d. Thus, the entitlement of the Employer to the payment of delay damages is negated for any circumstances which entitle the Contractor '*to an extension of the Time for Completion*'. As is demonstrated herein, the Contractor is entitled to such an extension of time and therefore the Employer is not entitled to the payment of delay damages by the Contractor.

4. Conditions Precedent to Entitlement

a. Sub-Clause 20.1 [Contractor's Claims] provides that:

> *If the Contractor considers himself to be entitled to any extension of the Time for Completion and/or any additional payment, under any Clause of these Conditions or otherwise in connection with the Contract, the Contractor shall give notice to the Engineer, describing the event or circumstance giving rise to the claim. The notice shall be given as soon as practicable, and not later than 28 days after the Contractor became aware, or should have become aware, of the event or circumstance.*
>
> *If the Contractor fails to give notice of a claim within such period of 28 days, the Time for Completion shall not be extended, the Contractor shall not be entitled to additional payment, and the Employer shall be discharged from all liability in connection with the claim. Otherwise, the following provisions of this Sub-Clause shall apply.*
>
> *The Contractor shall also submit any other notices which are required by the Contract and supporting particulars for the claim, all as relevant to such event or circumstance.*
>
> *The Contractor shall keep such contemporary records as may be necessary to substantiate any claim, either on the Site or at another location acceptable to the Engineer. Without admitting the Employer's liability, the Engineer may, after receiving any notice under this Sub-Clause, monitor the record-keeping and/ or instruct the Contractor to keep further contemporary records. The Contractor shall permit the Engineer to inspect all these records, and shall (if instructed) submit copies to the Engineer.*
>
> *Within 42 days after the Contractor became aware (or should have become aware) of the event or circumstance giving rise to the claim, or within such other period as may be proposed by the Contractor and approved by the Engineer, the Contractor shall send to the Engineer a fully detailed claim which includes full supporting particulars of the basis of the claim and of the extension of time and/or additional payment claimed.*

b. The requirement to give notice of entitlement to an extension of time and to describe the event giving rise to the claim within 28 days of the event is a condition precedent to the Contractor's entitlement. The Contractor is also obliged to submit a fully detailed claim of the extension of time claimed within 42 days of the event.

    c. The Contractor submitted a notice of claim on 15 February 2010[9] which is within the 28-day period prescribed in the Contract. The submission contained herein comprises the detailed claim and supporting particulars thereby satisfying the provisions of this sub-clause.

    d. The Contractor has therefore complied with the Conditions of this sub-clause and is consequently entitled to an extension of the Time for Completion until 6 August 2010.

5. Conclusion

    a. The following is a summary of the Contractor's entitlement as discussed in this section:

        i. The Contractor is entitled under the Contract to an extension to the Time for Completion for delay, impediment or prevention caused by or attributable to the Employer's other contractors on the Site.

        ii. Due to the circumstances entitling the Contractor to an extension of the Time for Completion, the Contractor is also entitled, pursuant to both the Contract and the law, to additional payment for the costs incurred as a result of the additional time he has been obliged to remain on site. The Contractor's claim in this respect will be submitted by way of a separate claim.

        iii. The Employer is not entitled to the payment of delay damages by the Contractor.

        iv. The Contractor has complied with the conditions precedent to entitlement.

---

[9] Exhibit 5 – BBO Construction letter reference BBOC/P9921/L0987, dated 15/02/10

The above narrative now comprises a robust claim which contains the essential elements of Cause, Effect, Entitlement and Substantiation (CEES). While the principles discussed here contain the *essential* elements of the claim, there are also several other issues that need to be addressed to provide a complete claim document and to make the reviewer's job easier and more pleasant; these will be discussed in the next chapter.

Summary of the Principles Covered in this Chapter

1. The object of a claim is to demonstrate that the claimant has entitlement to compensation and also to substantiate the amount of such compensation.
2. The object of a response document is to set out the findings of the determination of the claim.
3. The Cause is the event which has given rise to the claim and must be established and substantiated in the claim.
4. The Effect is a demonstration of how the event affected the claimant by linking the cause with the effect. This could be in terms of time or money and must be substantiated in the claim.
5. Entitlement is the claimant's right under the Contract or at law to the compensation claimed and must also be established and substantiated in the claim.
6. Conditions precedent to entitlement should be examined in the claim and it must be demonstrated that the claimant has complied with such conditions. Alternatively, if the claimant has not complied, a reasoned case must be made as to why the conditions should not affect the claimant's entitlement.
7. All statements, calculations and demonstrations must be substantiated by reference to the project records, the Contract or other such supporting evidence.
8. It is essential to establish robust contract administration systems to protect the claimant's contractual rights and to provide adequate substantiation of cause, effect and entitlement.
9. The essential elements to include in any claim or determination are CEES:
   • Cause
   • Effect
   • Entitlement
   • Substantiation

Chapter 5

Chapter 6

The Preliminaries to the Claim

Introduction

In previous chapters we have discussed the different types of claim that may arise on a construction contract, the presentation of a claim document, the importance of making the narrative readable, the reasons for making the submission a stand-alone document and the use and inclusion of the project records. We have also discussed at length the essential elements of cause, effect, entitlement and substantiation (CEES). While the treatment of cause, effect, entitlement and substantiation may be regarded as the *essential* elements to prove entitlement, there are several other issues that need to be addressed in order to fulfil the requirements already discussed, particularly making the claim submission a stand-alone document and leading the reviewer to a logical conclusion. Much of this can be achieved by working through the claim and presenting the various subjects that need to be dealt with in a logical order. In order to ensure that the claim is presented in such a way that it leads a reviewer through all the information in a logical manner, the presentation of a typical claim-submission document should be similar to the following:

1. Front Cover
2. Executive Summary
3. Statement of Claim
4. Definitions, Abbreviations and Clarifications
5. The Contract Particulars
6. The Method of Delay Analysis
7. Details of the Claim for an Extension of the Time for Completion
8. Details of the Claim for Additional Payment
9. Appendices
    a. Appendix A – Exhibits
    b. Appendix B – Baseline Programme
    c. Appendix C – Impacted As-Planned Programme
    d. Appendix D – Cost Calculations
    e. Appendix E – Supporting Information for the Cost Calculations

Construction Claims & Responses: Effective Writing & Presentation, Second Edition. Andy Hewitt.
© 2016 John Wiley & Sons, Ltd. Published 2016 by John Wiley & Sons, Ltd.

While the above list of contents is fairly typical, it is also true to say that there is no 'one size fits all' solution. For example, a claim for additional payment arising from a variation would be set out in a different way from a claim for an extension of time. Consequently, the contents of each section would be different, and the way of dealing with cause, effect, entitlement and substantiation would also be quite different in the two cases.

During a period when I was producing claims on a very regular basis, I developed a claim template whereby the 'skeleton' of the document was already laid out and certain passages of standard wording were included in the appropriate sections, leaving it to me to 'fill in the blanks' and flesh out the template according to the particular issue in question. While this was never intended to be a 'tick the boxes' solution to claims writing, this methodology was intended to make my life easier and my work more efficient. While this proved to be the case to some extent, I would say that on the majority of occasions it was necessary to amend this template to suit the circumstances of the particular issue and the same will be true for the example shown herein. In other words, the example shown here should be used more as a guide than as something that is set in stone.

In the remainder of this chapter, and in Chapters 7–9, we will examine a typical example of a claim for an extension of time and additional payment. We will work though this section by section by way of a discussion of the issues that should be included in each section and how these should be dealt with. We will also include an example of the claim narrative for each section in which the topics discussed will be put into practice. We will conclude each claim section with a checklist which may be used as a quality check for each section.

The Front Cover

Despite the old saying that you should never judge a book by its cover, given the fact that the front cover of a document is the part that is seen the most often, it is surprising that in many cases the front covers of important claim documents seem to have been put together as an afterthought. They do not contain fairly basic but important information and certainly do not invite the reviewer to look inside them in any way. A front cover of a claim document should contain the following as a minimum requirement:

1. The claimant's name
2. The project title
3. The claim title or brief description of the issue
4. The revision reference
5. The revision date

Additionally, the following information may be either desirable or necessary according to the particular situation:

1. The claimant's company logo
2. The other party's name

3. A rendering of the project
4. The claim number
5. The document reference number
6. The author
7. The reviewer
8. Graphics to make it look attractive

The inclusion of a document revision box is useful because in many cases a claim document will go through more than one version, both at the drafting stage and as the review, response and negotiations proceed. Consequently, there may be several versions in existence, so it is important to keep a track on revisions during the drafting stage and to clearly identify the appropriate revision once the document has been finalised and issued. For this reason, it is also sensible to include the revision number in the headers or footers of each page.

To use the analogy of the importance of creating a good impression when attending an important interview, the front cover is the equivalent of wearing a good suit and having polished shoes. In other words, the front cover will immediately provide an insight into the contents of the document. It is therefore important to ensure that it is well presented by the effective use of images, fonts, graphics, layout and formatting tools. Here is how the front cover could be presented for our example claim:

Please note that the example shown below contains only the basic information to allow for reproduction herein.

JOHNSON CONSTRUCTION GROUP

RED ROSE TOWER, NEWTOWN
CLAIM No. 6
for
An Extension to the Time for Completion and Additional Payment
in connection with
Delays Arising from the Electrical Transformer Room
Volume 1 of 2

Revision	Date	Submitted To	For	Prepared By	Reviewed By
0	19/04/13	Dawson-Wilkinson Partnership	The Engineer's Determination	C Woodward	S Thompson

Checklist – Front cover

1. Claimant's name
2. Project title
3. Claim title or brief description
4. Revision reference
5. Revision date
6. Company logo
7. The party's name
8. Rendering or picture of the project
9. Claim number
10. Volume number
11. Document reference number
12. Author
13. Reviewer
14. Attractive layout

Formatting of the Document

Before we move into the example claim, let's take a few moments to discuss the formatting of the actual document to ensure that it is presented professionally and is user-friendly as a working document.

The font should be such that it is easily readable, which generally means that artistic-type fonts should be avoided and that the font size should be either 11 or 12 point.

The page layout should be such that the left-hand side of the narrative does not 'disappear' when the document is bound. It is also a good idea to leave a reasonably sized right-hand margin so that there is space to make handwritten notes while reading or reviewing. I find that 25 mm (1 inch) margins are about right.

Line spacing can also help for both ease of reading and again for the making of notes or highlighting certain passages for emphasis or for later attention. I find that 1.5 or 2.0 line spacing works well in this respect.

Headers and footers may used effectively, possibly to identify the company that 'owns' the document, the title of the document, the revision number, the date of the revision and the page numbers. The layout and fonts of headers and footers may be used effectively to enhance the appearance of the document.

Typical examples of headers and footers are as follows:

Chapter 6

Johnson Construction Group
Red Rose Tower

Narrative	
Narrative	
Narrative	

Claim No. 6	*Page 5 of 98*	*Rev. 0, 19/04/13*

Formatting may occasionally be lost during production of a document, especially when cutting and pasting sections between documents. It is therefore a good idea to perform a quality check on completion, to ensure that formatting, fonts, layouts and the like remain consistent throughout.

The Contents

The list of contents should appear on the first page following the front cover. While this should be fairly obvious, many documents omit this simple item with the result that the reviewer can spend needless time when trying to locate a particular section or item. A typical layout would be as follows:

Chapter 6

CONTENTS

Section		Page
Section 1	Executive Summary	: 3
Section 2	Statement of Claim	: 5
Section 3	Definitions, Abbreviations and Clarifications	: 8
Section 4	The Contract Particulars	: 10
Section 5	The Method of Delay Analysis	: 15
Section 6	Details of the Claim for an Extension of the Time for Completion	: 20

Checklist – Contents

1. Section Numbers.
2. Section titles.
3. Page numbers.

The Contents could, of course, be expanded to include subsections. This is particularly useful if the document is lengthy or includes several heads of claim.

The Executive Summary

While there may be several people who need to review the claim document in detail, it is also true that there will be others who need only a distillation of the contents of the document. It is also useful for those who will be involved at a detailed level to gain an overview of, and insight into, the issues that are to be dealt with and thus get a feel for the matter before they proceed to the details. An executive summary is essentially a summary of the entire contents, distilled down to just a few pages and it provides the reader the opportunity to obtain an understanding of the whole issue very quickly.

It should be obvious that the executive summary cannot be drafted until all the other sections have been completed. Somewhat illogically, therefore, in terms of how we proceed to draft the claim document, this needs to be the first part of the narrative and should follow immediately after the Contents page, but will be the last section to be written. As we will see later, if the various sections of the claim are approached in a logical manner, the executive summary may be fairly easily compiled from the introductions and conclusions of the individual sections. I find that it is good practice to leave the drafting of the executive summary until internal reviews have been carried out, because this ensures that it accurately reflects the final version that will be 'locked down' for submission.

An executive summary for our example claim could be as follows:

SECTION 1
EXECUTIVE SUMMARY

1. A Contract was entered into between Jason Leonard Developments (hereinafter called 'the Employer' or 'JLD') and Johnson Construction Group (hereinafter called 'the Contractor' or 'JCG') for the construction of a mixed-use, 16-storey building known as Red Rose Tower which includes heating, ventilating and air-conditioning systems throughout. The building is designed to a high technical specification, particularly with regard to energy conservation, and a sophisticated building-management system is included.

2. The Contractor asserts that delays arising out of a redesign of the Electrical Transformer Room caused a delay for which he is entitled to an extension of the Time for Completion.

3. The Contractor also considers that he is entitled to additional payment to compensate him for the additional time that he was obliged to remain on site and to provide contributions to his head-office running costs.

4. The Contractor has completed a delay analysis to demonstrate the extension of time to which he considers himself entitled, a detailed explanation of which is included herein under Section 5. This explains that the latest as-planned programme has been used to create an impacted as-planned programme in order to demonstrate the effect of the delay event on the Time for Completion.

5. Section 6 herein demonstrates the following:
   a. The requirements of the National Electricity Company resulted in the Engineer being obliged to instruct the Contractor to carry out alterations to the already constructed Electrical Transformer Room to revise the layout in order to meet these requirements.
   b. The instruction for this work was issued to the Contractor at a time when the work affected the programmed dates for installation and commissioning of the electrical transformers, the subsequent power-on date and the start of the testing and commissioning activities.
   c. The Contractor took various steps to minimise the time taken to carry out the alteration work and also mitigated the delay by completing the testing and commissioning activity within a reduced duration from that which was planned.
   d. The Contractor has demonstrated the effect of the delay by producing a delay analysis which demonstrates that the instructed work caused a delay to the completion of the project of 30 days, resulting in a revised completion date of 11 March 2013.
   e. The Contractor has demonstrated that there were no Contractor-caused delays concurrent with this delay event.
   f. The Contractor's entitlement to an extension of the Time for Completion is contained in the following clauses:
      i. Sub-Clause 1.9 [*Delayed Drawings or Instructions*]
      ii. Sub-Clause 4.7 [*Setting Out*]
      iii. Sub-Clause 8.4 [*Extension of Time for Completion*]
      iv. Sub-Clause 8.5 [*Delays Caused by Authorities*]
   g. The Contractor has complied with the provisions of Sub-Clause 20.1 [*Contractor's Claims*] with regard to the issue of a notice of claim and the submission of his claim and detailed particulars to a fair and reasonable extent.
   h. The reasons entitling the Contractor to an extension of the Time for Completion for the delay have negated the Employer's entitlement to the payment of delay damages.

6. Section 7 herein demonstrates the following:
   a. The Contractor's entitlement to additional payment is contained in the following clauses:
      i. Sub-Clause 1.9 [*Delayed Drawings or Instructions*]
      ii. Sub-Clause 4.7 [*Setting Out*]
   b. The Contractor has complied with the provisions of Sub-Clause 20.1 [*Contractor's Claims*] to a fair and reasonable extent.

   c. The additional payment to which the Contractor is entitled
      consists of:
      i. Site Establishment Costs
      ii. Contractual Costs
      iii. Head-Office Overheads and Profit
      iv. Finance Costs
      v. Reasonable Profit
   d. Substantiation of the costs has been provided by way of docu-
      ments included herein under Appendix J.
7. The Contractor therefore claims an extension of the Time for
   Completion to 11 March 2013 for the delay event included in this
   claim and requests the Engineer, as is required by the Contract,
   to make a fair determination of such.
8. The Contractor also claims that, pursuant to the Contract and to
   common law, he should be awarded an additional payment of
   £331,045 to compensate him for additional costs expended
   throughout the period of prolongation, plus reasonable profit,
   and he similarly requests the Engineer to make a fair determination in this respect in accordance with the Contract.

Checklist – Executive summary

1. Contains a summary of all the Sections.
2. Includes any last-minute changes and revisions to the main
   narrative.

The Statement of Claim

Care must be taken not to just repeat the contents of the executive summary within the statement of claim. The two sections are included for
separate purposes. The executive summary should provide the reviewer
with a condensed version of the full claim, and the statement of claim
should briefly set the scene for the remainder of the document by providing brief details of the contract, the project, the nature of the claim and
the circumstances giving rise to the claim. Cause, effect and entitlement
should also be touched on here and we should also set out both the
Contractor's and the Employer's obligations with regard to the administration of the claim. This section should also state the quantum of the claim and

Chapter 6

whether the claim is an interim or a final submission. It should also be stated whether the document is a resubmission or a revision to the claim.

From the following example, you will notice that the 'tone' of the writing is established here by references to the contractual language and by referring to the parties by the same definitions as used in the Contract. We are also abiding by the rule that the document should be a stand-alone document that may be understood by someone with no familiarity with the project, by including the details of the Contract and the parties in the opening statement.

SECTION 2
STATEMENT OF CLAIM

Description of Claim

1. A Contract agreement (hereinafter called 'the Contract') was entered into on 16 December 2010 between Jason Leonard Developments (hereinafter called 'the Employer' or 'JLD') and Johnson Construction Group (hereinafter called 'the Contractor' or 'JCG') for the construction of a mixed-use, 16-storey building known as Red Rose Tower (hereinafter called 'the Project').

2. The Project includes an Electrical Transformer Room on the ground floor in which the transformers were to be located. In January 2013, it became apparent that the Electrical Transformer Room, which had already been constructed by the Contractor in accordance with the issued-for-construction drawings, did not comply with the requirements of the local electricity authority. The Employer was obliged to redesign the area in question and, consequently, the Contractor was instructed by the Engineer to carry out demolition and alteration works and to reconstruct substantial elements of the room. The time taken for consultation, redesign and for the construction works to be carried out delayed the installation of the transformers and thus the power-on date.

3. The Project is designed to a high technical specification and includes air-conditioning throughout and a sophisticated building-management system to reduce energy consumption. The delay in the provision of 'clean' electrical power by way of the main electricity connection delayed the testing and commissioning of the mechanical, electrical and building-management systems and, consequently, the final completion of the project.

4. The Contract provides that in circumstances where the Employer instructs a Variation to the Contract, or where delays are caused

by the late issue of drawings and/or instructions, or by public authorities, the Engineer is obliged to determine the amount of the extension of the Time for Completion of the Works. This claim therefore comprises the Contractor's detailed claim and supporting particulars, to enable the Engineer to determine the extension of the Time for Completion of the Works.

5. Due to the delay to the Time for Completion, the Contractor has also been obliged to maintain his site establishment for a period longer than contemplated had the Contract been performed as originally envisaged and has consequently incurred additional costs in doing so. The Contractor therefore considers that he is entitled to an award of an additional payment to enable him to recover such costs. This claim therefore also contains the Contractor's claim for additional costs for prolongation associated with the extension of the Time for Completion and includes supporting particulars to enable the Engineer to determine the amount of the additional payment.

6. The Contractor hereby claims an extension of the Time for Completion until 11 March 2013 and an additional payment of £331,045 for prolongation costs that have been incurred due to the extension of time.

7. The events giving rise to the extension of the Time for Completion of the Works have been finalised and this claim is therefore the Contractor's entire claim for this event.

Chapter 6

Checklist – Statement of claim

1. Brief details of the Contract
2. Details of the parties
3. Brief details and description of the Project
4. Brief details of the events giving rise to the claim
5. Brief details of entitlement
6. Details of the Contract procedure for claim submissions and determination
7. Amount(s) claimed
8. States whether a final or interim claim

Definitions, Abbreviations and Clarifications

You will probably have noticed that one of the first things to be stated in any form of contract is a list of definitions. Such a list defines the meanings of certain names or phrases used in the Contract. It is therefore good practice to provide consistency between the claim document and the Contract by using the same names and phrases in the claim as are used in the Contract. For example, if the Contract refers to the 'Time for Completion' as does FIDIC, then it may be confusing to refer to the 'completion date'. It is good practice to repeat such contractual definitions in the claim. While a good claim document will maintain wording consistent with the contractual language and avoid the use of abbreviations, acronyms and project-specific references, it is quite probable that the use of certain project-specific abbreviations, acronyms and definitions will become inevitable through references included in the drawings and specifications, or from usage in correspondence, reports and the like. It is therefore also important to include definitions of such terms at this point, so that a person who is unfamiliar with the project is made aware of their meaning.

Similarly, it is also sensible to inform the reviewer how the claim document is set out and presented so that he knows, for example, whether he is looking at a quotation or a paraphrase and how the narrative is cross-referenced to the substantiating documents. A typical section in relation to our example claim is as follows:

SECTION 3
DEFINITIONS, ABBREVIATIONS
AND CLARIFICATIONS

Introduction

1. This Section includes a list of definitions, abbreviations and clarifications included in both the Contract and this claim document.
2. This Section also explains the format and arrangement of the claim document.

Definitions

1. 'Employer' or 'JLD' means Jason Leonard Developments.
2. 'Contractor' or 'JCG' means Johnson Construction Group.
3. 'Engineer' or 'DWP' means Dawson-Wilkinson Partnership, the consultants appointed by the Employer to act as the Engineer under the Contract.
4. 'Corrie Design and Engineering' or 'CDE' means the Employer's designers of the Project.
5. 'National Electricity Company' or 'NEC' means the local authority providing electrical power to the Project.

6. 'Contract' means the Contract Agreement, the Letter of Acceptance, the Letter of Tender, the Conditions, the Specification, the Drawings, the Schedules, and the further documents that are listed in the Contract Agreement.

7. 'FIDIC' means the Fédération Internationale des Ingénieurs-Conseils, the International Federation of Consulting Engineers.

8. 'Time for Completion' means the time for completing the Works or a Section (as the case may be) under Sub-Clause 8.2 [*Time for Completion*], as stated in the Appendix to Tender [with any extension under Sub-Clause 8.4 [*Extension of Time for Completion*]), calculated from the Commencement Date.

9. 'Day' means a calendar day and 'year' means 365 days.

10. 'Contract Price' means the price defined in Sub-Clause 14.1 [*The Contract Price*], and includes adjustments in accordance with the Contract.

11. 'Cost' means all expenditure reasonably incurred (or to be incurred) by the Contractor, whether on or off the Site, including overhead and similar charges, but does not include profit.

12. 'Contractor's Equipment' means all apparatus, machinery, vehicles and other things required for the execution and completion of the Works and the remedying of any defects. However, Contractor's Equipment excludes Temporary Works, Employer's Equipment (if any), Plant, Materials and any other things intended to form or forming part of the Permanent Works.

13. 'Works' mean the Permanent Works and the Temporary Works, or either of them as appropriate.

14. 'Site' means the places where the Permanent Works are to be executed and to which Plant and Materials are to be delivered and any other places as may be specified in the Contract as forming part of the Site.

15. 'Variation' means any change to the Works that is instructed or approved as a Variation under Clause 13 [*Variations and Adjustments*].

Format and Arrangement of the Claim Document

1. Quotations from project documents and other sources are shown in quotation marks and italics '*thus*'.

2. Where references are made to project records and other documents in this narrative, the relevant document is included herein in Appendix A for verification. For ease of reference, the documents included in Appendix A are presented in chronological order and are cross-referenced with footnotes, thus.[1]

---

[1] Example of footnote cross-reference

Checklist – Definitions, abbreviations and clarifications

1. Introduction – the purpose of this Section
2. Definitions and abbreviations of the parties
3. Contractual definitions
4. Method of dealing with quotations
5. Method of dealing with cross-references
6. Arrangement of the claim document

The Contract Particulars

This section will deal specifically with the conditions of contract. Remembering the importance of creating a document that someone with no prior knowledge of the project can understand, it is very useful for the reviewer at this stage to know whether he is dealing with a claim on a road-construction project, a process plant or a new shopping mall. Consequently, the first thing to start with is a description of the Works so that the reviewer will immediately get a feeling for the project and the likely conditions and circumstances of the claim.

The parties should also be introduced at this stage and their roles and responsibilities under or outside the Contract explained. In addition to the Employer and the Contractor, this would typically include the Engineer, the designer and the cost consultant.

The next issue to deal with at this stage is the Contract itself, with a brief outline of the conditions. Is this a standard form of contract such as FIDIC? If so, the particular form, edition and date should be stated. If the conditions are based upon a standard form with separate particular conditions, this should also be noted here. Possibly it is the Employer's own bespoke contract and, in such a case, a brief explanation should be included here. In some areas, particularly in the case of international contracts, the applicable law may not be that of the country where the project is located and, in such circumstances, it is a good idea to clarify the applicable law at this point.

If the Contract is based upon a standard form of contract plus particular conditions, care must be taken to be aware of, and to take into account, any amendments to the standard clauses that have been made within the particular conditions. I was once obliged to reject a contractor's claim entirely, because the contractor had not taken the trouble to study the particular conditions and this affected the whole basis of his claim. A good way to ensure that such errors cannot be made is to make a working copy of the contract document in which the particular conditions are consolidated into the standard form. If a soft copy of the standard form is available, this may be fairly easily accomplished by the use of the usual computer programs. Otherwise, copies of the particular conditions may be manually cut and pasted (and by manually, I mean with scissors and glue), either over the standard clause, or onto the facing page of the standard conditions. If this task is completed at the commencement of the project, I can assure you that as well as reducing the possibility of errors, it will save much time and frustration by the avoidance of having to leaf through two separate documents and to try and mentally consolidate the clauses.

Certain key details from the Contract will also be useful to the reviewer for background information, so it is necessary to extract those from the Contract and summarise them here. Such details would typically include the following:

1. The Tender Date
2. The Contract Sum
3. Dates of Commencement and Completion
4. Milestone dates

It could be the case that the Contractor has been awarded previous extensions of time and additional payment and, in such cases, it would be appropriate to include a brief description of these, in order that the reviewer may be apprised of the fact that the completion date and the contract sum have already been changed from those stated in the Contract.

We have discussed, in Chapter 5, the fact that one of the four essential elements of a successful claim is to establish the claimant's entitlement under the Contract. To achieve this, it is necessary to make reference to the contract conditions that provide entitlement for the event or events in question. We have also discussed the importance of making the reviewer's job as easy and as pleasant as possible, and that one of the ways in which we can achieve this goal is to make the claim a stand-alone document. It is therefore a good idea at this stage to reproduce the conditions of contract, which will be later relied on within the claim. In the real world of claims' writing, it is often not possible at this stage in the production of the claim to ascertain exactly which clauses it will eventually need to rely upon. This task can therefore be left until such time as the necessary research and examination of the Contract has been undertaken.

It may seem like an onerous task to reproduce vast sections of text from the Contract, especially when the reviewer should already have a copy of the Contract for reference. While admittedly such reproduction may take some time, this procedure does offer some distinct advantages. Most importantly perhaps is that the very act of typing out the clause word for word makes for a clear understanding of the meaning of the clause. You may have experienced a situation where you *thought* that you knew all the particular nuances of a clause, possibly based on what someone had previously told you, or based on a similar clause in another contract, but it was not until you had cause to examine the clause in detail that you realised that your interpretation was incorrect in some aspect or other. For

Chapter 6

this reason, unless you are sure that you know the clause inside out, upside down and backwards, the temptation to cut and paste the relevant clauses into the narrative, to have your secretary type them for you, or to photocopy them into an appendix should be resisted. In the case of a standard form of contract that has been supplemented or amended by the use of particular conditions, the particular conditions should be incorporated into the reproduced clause. This provides an easier and better understanding of the complete clause for both the writer and the reviewer. On a practical note, I have found it good practice that, whenever I need to reproduce a clause, I cut and paste the reproduced clause into a reference document and save it for future use. It is surprising how many times the same few clauses need to be used or referred to on a typical project, perhaps in letters, reports or in subsequent claims. This simple practice can save significant time over the life of a project.

The reproduction of the relevant clauses at this point in the claim document also helps to provide the reviewer with an insight into the events that will be presented later in the narrative, so that he has already got an idea of what he should have in mind to enable him to make his determination. It also deals with the possibility (and believe me, this does happen) that the reviewer does not have a copy of the contract or the particular conditions in his possession or, if he does have a copy of the contract, he is not aware that it should be read in conjunction with the particular conditions. If such a situation does arise, it could at best result in a delay in the response time or at worst an erroneous determination, which would then need to be challenged and lead to delay in the resolution of the matter. One must consider here how much extra time such a situation might require compared to the extra effort taken to produce a good claim document in the first place.

Let us now look at how we could deal with this section in our example claim:

SECTION 4
THE CONTRACT PARTICULARS

Introduction

This Section shows the relevant details from the Contract.

Project Description

1. The project was designed by the Employer and consists of three levels of basement parking built on piled foundations, ground-floor retail units, four floors of commercial office space and

twelve storeys of high-specification two and three-bedroom residential apartments.

2. The basement parking extends to the boundary of the site. The 17-storey tower block containing the retail, commercial and residential units occupies approximately 60% of the site plan area, and the remaining area at ground level above the basement consists of soft and hard landscaping.

3. The project includes heating, ventilating and air-conditioning systems throughout and is designed to a high technical specification, particularly with regard to energy conservation. A sophisticated building-management system is included.

4. With the exception of the communal areas, the retail and commercial areas are to be completed to shell and core stage for future tenant fit-out. The residential units include finishes and decoration.

5. The project is located within an inner-city 'brown field' development in the central area of Newtown.

Names of the Parties to the Contract and the Consultants

1. The parties involved in the Contract are as follows:
   a. The Employer: Jason Leonard Developments (JLD).
   b. The Contractor: Johnson Construction Group (JCG).
   c. The Engineer: Dawson-Wilkinson Partnership (DWP).

Details of the Conditions of Contract Governing the Contract and Applicable Law

1. The general conditions of contract are the *Conditions of Contract for Construction for Building and Engineering Works Designed by the Employer, First Edition 1999.*

2. The law to which the contract is subject is that of England and Wales.

The Tender Date

The Contractor's tender was submitted on 15 October 2010. Various revisions and amendments were made to the Contractor's offer and these were subsequently incorporated into the Contract under the Employer's Letter of Acceptance dated 29 November 2010[2] which was accepted by the Contractor on 3 December 2010[3].

The Contract Sum

The Contract sum as confirmed in the Letter of Acceptance is £50,182,513 on a lump sum basis.

---

[2] Exhibit 2 – JLD letter reference JLD/let/JCG/L-001, dated 29/11/10
[3] Exhibit 3 – JCG letter reference P-1013/Let-0001, dated 03/12/10

Chapter 6

Dates for Commencement and Completion

1. The Commencement Date was 3 January 2011 as confirmed in the Letter of Acceptance.
2. The Time for Completion was 31 December 2012 as confirmed in the Letter of Acceptance.

Previous Extensions of the Time for Completion and Additional Payment

1. The Contractor has submitted previous claims for extensions of the time for completion and the Engineer has made awards in this respect as follows:
   a. Extension of Time Award No. 1 for Claim No. 1 for Unforeseeable Physical Conditions which resulted in a substantial increase in the piling and foundation works and for which an extension of time until 21 January 2013 was awarded.[4]
   b. Extension of Time Award No. 2 for Claim No. 4 for the late nomination of the facade contractor and for which an extension of time until 9 February 2013 was awarded.[5]
2. The Contractor has been awarded the additional payment of £184,136 for prolongation costs in respect of Claim No. 1 described above.
3. The Contractor has submitted a claim for additional payment of £170,497 for prolongation costs in respect of Claim No. 4 described above, but to date negotiations have not been concluded and the Engineer's determination is still outstanding.
4. The Contractor has made several claims for additional payment as a result of variation instructions, the evaluation and agreement of which is currently ongoing.

Conditions of Contract

The conditions of contract that have relevance to this claim are reproduced below for ease of reference:

1. *'1.9 Delayed Drawings or Instructions*
   The Contractor shall give notice to the Engineer whenever the Works are likely to be delayed or disrupted if any necessary drawing or instruction is not issued to the Contractor within a particular time, which shall be reasonable. The notice shall include details of the necessary drawing or instruction, details of why and by when it should be issued, and details of the nature and amount of the delay or disruption likely to be suffered if it is late.

---

[4] Exhibit 6 – DWP letter reference ST/JW/Contr/126, dated 22/09/11
[5] Exhibit 7 – DWP letter reference ST/JW/Contr/295, dated 17/01/12

If the Contractor suffers delay and/or incurs Cost as a result of a failure of the Engineer to issue the notified drawing or instruction within a time which is reasonable and is specified in the notice with supporting details, the Contractor shall give a further notice to the Engineer and shall be entitled subject to Sub-Clause 20.1 [Contractor's Claims] to:

a. an extension of time for any such delay, if completion is or will be delayed, under Sub-Clause 8.4 [Extension of Time for Completion], and

b. payment of any such Cost plus reasonable profit, which shall be included in the Contract Price.

After receiving this further notice, the Engineer shall proceed in accordance with Sub-Clause 3.5 [Determinations] to agree or determine these matters.

However, if and to the extent that the Engineer's failure was caused by any error or delay by the Contractor, including an error in, or delay in the submission of, any of the Contractor's Documents, the Contractor shall not be entitled to such extension of time, Cost or profit.'

2. '3.3 Instructions of the Engineer

The Engineer may issue to the Contractor (at any time) instructions and additional or modified Drawings which may be necessary for the execution of the Works and the remedying of any defects, all in accordance with the Contract. The Contractor shall only take instructions from the Engineer, or from an assistant to whom the appropriate authority has been delegated under this Clause. If an instruction constitutes a Variation, Clause 13 [Variations and Adjustments] shall apply.

The Contractor shall comply with the instructions given by the Engineer or delegated assistant, on any matter related to the Contract. Whenever practicable, their instructions shall be given in writing. If the Engineer or a delegated assistant:

a. gives an oral instruction,

b. receives a written confirmation of the instruction, from (or on behalf of) the Contractor, within two working days after giving the instruction, and

c. does not reply by issuing a written rejection and/or instruction within two working days after receiving the confirmation,

then the confirmation shall constitute the written instruction of the Engineer or delegated assistant (as the case may be).'

3. '3.5 Determinations

Whenever these Conditions provide that the Engineer shall proceed in accordance with this Sub-Clause 3.5 to agree or determine any matter, the Engineer shall consult with each Party in an endeavour to reach agreement. If agreement is not achieved, the Engineer *shall make a fair determination in accordance with the Contract, taking due regard of all relevant circumstances.*

The Engineer shall give notice to both Parties of each agreement or determination, with supporting particulars. Each Party shall give effect to each agreement or determination unless and until revised under Clause 20 [Claims, Disputes and Arbitration].'

4. *'4.7 Setting Out*

The Contractor shall set out the Works in relation to original points, lines and levels of reference specified in the Contract or notified by the Engineer. The Contractor shall be responsible for the correct positioning of all parts of the Works, and shall rectify any error in the position, levels, dimensions or alignment of the Works.

The Employer shall be responsible for any errors in these specified or notified items of reference, but the Contractor shall use reasonable efforts to verify their accuracy before they are used.

If the Contractor suffers delay and/or incurs Cost from executing work which was necessitated by an error in these items of reference, and an experienced contractor could not reasonably have discovered such error and avoided this delay and/or Cost, the Contractor shall give notice to the Engineer and shall be entitled subject to Sub-Clause 20.1 [Contractor's Claims] *to:*

a. *an extension of time for any such delay, if completion is or will be delayed, under Sub-Clause 8.4* [Extension of Time for Completion], *and*

b. *payment of any such Cost plus reasonable profit, which shall be included in the Contract Price.*

After receiving this notice, the Engineer shall proceed in accordance with Sub-Clause 3.5 [Determinations] *to agree or determine (i) whether and (if so) to what extent the error could not reasonably have been discovered, and (ii) the matters described in subparagraphs (a) and (b) above related to this extent.'*

5. *'8.4 Extension of Time for Completion*

The Contractor shall be entitled subject to Sub-Clause 20.1 [Contractor's Claims] *to an extension of the Time for Completion if and to the extent that completion for the purposes of Sub-Clause 10.1* [Taking Over of the Works and Sections] *is or will be delayed by any of the following causes:*

a. *a Variation (unless an adjustment to the Time for Completion has been agreed under Sub-Clause 13.3* [Variation Procedure]*) or other substantial change in the quantity of an item of work included in the Contract,*

b. *a cause of delay giving an entitlement to extension of time under a Sub-Clause of these Conditions,*

c. *exceptionally adverse climatic conditions,*

Chapter 6

    d. Unforeseeable shortages in the availability of personnel or Goods caused by epidemic or governmental actions, or

    e. any delay, impediment or prevention caused by or attributable to the Employer, the Employer's Personnel, or the Employer's other contractors on the Site.

If the Contractor considers himself to be entitled to an extension of the Time for Completion, the Contractor shall give notice to the Engineer in accordance with Sub-Clause 20.1 [Contractor's Claims]. *When determining each extension of time under Sub-Clause 20.1, the Engineer shall review previous determinations and may increase, but shall not decrease, the total extension of time.'*

6. *'8.5 Delays Caused by Authorities*
   If the following conditions apply, namely:
       a. the Contractor has diligently followed the procedures laid down by the relevant legally constituted public authorities in the Country,
       b. these authorities delay or disrupt the Contractor's work, and
       c. the delay or disruption was Unforeseeable,
   then this delay or disruption will be considered as a cause of delay under sub-paragraph (b) of Sub-Clause 8.4 [Extension of Time for Completion].'

7. *'8.7 Delay Damages*
   If the Contractor fails to comply with Sub-Clause 8.2 [Time for Completion], *the Contractorshall subject to Sub-Clause 2.5* [Employer's Claims] *pay delay damages to the Employer for this default. These delay damages shall be the sum stated in the Appendix to Tender, which shall be paid for every day which shall elapse between the relevant Time for Completion and the date stated in the Taking-Over Certificate. However, the total amount due under this Sub-Clause shall not exceed the maximum amount of delay damages (if any) stated in the Appendix to Tender.*

   These delay damages shall be the only damages due from the Contractor for such default, other than in the event of termination under Sub-Clause 15.2 [Termination by Employer] *prior to completion of the Works. These damages shallnot relieve the Contractor from his obligation to complete the Works, or from any other duties, obligations or responsibilities which he may have under the Contract.'*

8. *'13.1 Right to Vary*
   Variations may be initiated by the Engineer at any time prior to issuing the Taking-Over Certificate for the Works, either by an instruction or by a request for the Contractor to submit a proposal.

The Contractor shall execute and be bound by each Variation, unless the Contractor promptly gives notice to the Engineer stating (with supporting particulars) that the Contractor cannot readily obtain the Goods required for the Variation. Upon receiving this notice, the Engineer shall cancel, confirm or vary the instruction.

Each Variation may include:

a. *changes to the quantities of any item of work included in the Contract (however, such changes do not necessarily constitute a Variation),*

b. *changes to the quality and other characteristics of any item of work,*

c. *changes to the levels, positions and/or dimensions of any part of the Works,*

d. *omission of any work unless it is to be carried out by others,*

e. *any additional work, Plant, Materials or services necessary for the Permanent Works, including any associated Tests on Completion, boreholes and other testing and exploratory work, or*

f. *changes to the sequence or timing of the execution of the Works.*

The Contractor shall not make any alteration and/or modification of the Permanent Works, unless and until the Engineer instructs or approves a Variation.'

9. *'20.1 Contractor's Claims*

If the Contractor considers himself to be entitled to any extension of the Time for Completion and/or any additional payment, under any Clause of these Conditions or otherwise in connection with the Contract, the Contractor shall give notice to the Engineer, describing the event or circumstance giving rise to the claim. The notice shall be given as soon as practicable, and not later than 28 days after the Contractor became aware, or should have become aware, of the event or circumstance.

If the Contractor fails to give notice of a claim within such period of 28 days, the Time for Completion shall not be extended, the Contractor shall not be entitled to additional payment, and the Employer shall be discharged from all liability in connection with the claim. Otherwise, the following provisions of this Sub-Clause shall apply.

The Contractor shall also submit any other notices which are required by the Contract, and supporting particulars for the claim, all as relevant to such event or circumstance.

The Contractor shall keep such contemporary records as may be necessary to substantiate any claim, either on the Site or at another location acceptable to the Engineer. Without admitting the Employer's liability, the Engineer may, after receiving any notice under this Sub-Clause, monitor the record-keeping and/or instruct the Contractor to keep further contemporary records. The Contractor shall permit the Engineer to inspect all these records, and shall (if instructed) submit copies to the Engineer.

Within 42 days after the Contractor became aware (or should have become aware) of the event or circumstance giving rise to the claim, or within such other period as may be proposed by the Contractor and approved by the Engineer, the Contractor shall send to the Engineer a fully detailed claim which includes full supporting particulars of the basis of the claim and of the extension of time and/or additional payment claimed. If the event or circumstance giving rise to the claim has a continuing effect:

a. *this fully detailed claim shall be considered as interim;*
b. *the Contractor shall send further interim claims at monthly intervals, giving the accumulated delay and/or amount claimed, and such further particulars as the Engineer may reasonably require; and*
c. *the Contractor shall send a final claim within 28 days after the end of the effects resulting from the event or circumstance, or within such other period as may be proposed by the Contractor and approved by the Engineer.*

Within 42 days after receiving a claim or any further particulars supporting a previous claim, or within such other period as may be proposed by the Engineer and approved by the Contractor, the Engineer shall respond with approval, or with disapproval and detailed comments. He may also request any necessary further particulars, but shall nevertheless give his response on the principles of the claim within such time.

Each Payment Certificate shall include such amounts for any claim as have been reasonably substantiated as due under the relevant provision of the Contract. Unless and until the particulars supplied are sufficient to substantiate the whole of the claim, the Contractor shall only be entitled to payment for such part of the claim as he has been able to substantiate.

The Engineer shall proceed in accordance with Sub-Clause 3.5 [Determinations] *to agree or determine (i) the extension (if any) of the Time for Completion (before or after its expiry) in accordance with Sub-Clause 8.4* [Extension of Time for Completion], *and/or (ii) the additional payment (if any) to which the Contractor is entitled under the Contract.*

The requirements of this Sub-Clause are in addition to those of any other Sub-Clause which may apply to a claim. If the Contractor fails to comply with this or another Sub-Clause in relation to any claim, any extension of time and/or additional payment shall take account of the extent (if any) to which the failure has prevented or prejudiced proper investigation of the claim, unless the claim is excluded under the second paragraph of this Sub-Clause.'

Chapter 6

Checklist – The Contract particulars

1. Introduction – the purpose of this section
2. Details of the parties – the Employer
3. Details of the parties – the Contractor
4. Details of the parties – the Engineer
5. Details of the parties – other relevant parties
6. The form of contract
7. The applicable law
8. The Tender Date
9. The Contract sum
10. The Commencement date
11. The Completion date
12. Previous extension-of-time awards
13. Previous awards for additional payment
14. Milestone dates
15. The relevant conditions of contract

Chapter 7

The Extension of Time Claim

Thus far, the claim narrative has primarily been concerned with setting the scene and providing the reviewer with background information. Let's take a moment to review what has been covered so far.

1. A description of the project
2. The details of the parties
3. The details of the contract particulars
4. Brief details of the delay event upon which the claim is based
5. Definitions used in the Contract and in the claim document
6. The contractual clauses that will be referenced in the claim
7. Details of previous claims and awards

You may think that so far we have done an awful lot of work without going into many of the details and circumstances surrounding the actual delay event itself and you may also consider that such important details should be covered much earlier in the narrative. Let us however, put ourselves in the place of a reviewer and ask whether or not we would prefer to understand the background of the project and have basic information such as whether it is a pipeline project or a housing development before or after we are presented with the details of the event. Would it not also be useful to us to be aware that when a document refers to JCG, for example, this means the Contractor and not the cost consultant? Let us also bear in mind that if there is no entitlement under the contract, then there is no basis for a claim, so it would be useful to know on what contractual basis the claim is made before going into the details. It is for these reasons that it is important to set the scene before we proceed to the main subject matter, which will be dealt with in the remainder of this chapter and Chapter 8.

Construction Claims & Responses: Effective Writing & Presentation, Second Edition. Andy Hewitt.
© 2016 John Wiley & Sons, Ltd. Published 2016 by John Wiley & Sons, Ltd.

The Method of Delay Analysis

An essential part of any claim for an extension of time is a demonstration that the delay event actually had an effect on the time for completion. It is not the purpose of this book to delve into the often murky science of delay analysis, so I would like at this point to recommend the Society of Construction Law's *Delay and Disruption Protocol* which goes a long way towards explaining the principles and making recommendations as to the most suitable forms of project programming and delay analyses. Certain principles, however, do need to be understood and taken into account in order to demonstrate entitlement to additional time, so it is worthwhile overviewing the basics here.

Firstly, there must be a programme from which to measure the effect of a delay. This may be the Contractor's agreed programme, or it may be a later programme that has been updated to take into account previous extensions of time and shows a revised date for completion. Whatever programme is used for the delay analysis, it is necessary to make reference to it within the claim, possibly with an explanation as to how it came about. It is also necessary to include substantiation to demonstrate that it was approved or accepted by the Engineer and to include the programme within the appendices.

It must be remembered that delay does not automatically lead to an extension of time. For example, a 25-day delay to an activity will not automatically result in entitlement to an extension of time of 25 days because, in order to have an effect on the completion date, the delay event must impact the critical path of the programme. It could well be the case that a delay event will only use up float within the programme and will have no effect on the completion date, or that the delayed activities are not on the critical path. Alternatively, it could be the case that the delay event uses up the entire available float, thus making the activity part of the critical path and therefore affecting the date for completion. In the latter case, however, the 25-day delay would affect the completion date by a lesser number of days, as shown in the following example:

Number of days delay:	25 days
LESS: Available float:	16 days
Effect on the completion date:	9 days

While the above case relies on the principle that the project owns the float, there are also some arguments to the effect that, because the programme is 'owned' by the Contractor, the Contractor also owns the float. Following this point of view, it is argued that the float should be preserved in any extension-of-time claims, in order to retain a future provision against contractor-caused delays.

I once spent a considerable number of hours preparing a claim document on behalf of a contractor after being advised by the contractor's project manager that a certain delay event justified an extension of time. It was only after I had done most of my work and the delay event was impacted into the baseline programme by the contractor's planner, that it was discovered that there was actually no effect on the completion date and, consequently, my client was not actually entitled to an extension of time at all. For this reason, it is important to run the delay analysis as soon as the circumstances causing the delay have ceased, so that the effect, if any, on the completion date may be seen.

One of the frequently used methods of delay analysis is to impact the delay into the current programme to produce an 'impacted as-planned' programme, and this may be done by several methods according to the circumstances. Some examples are as follows:

1. If the delay event affects the start date of an activity, then the planned start date should be adjusted accordingly.
2. Possibly the delay event prevented the end date of an activity from being achieved, in which case the end date should be adjusted.
3. The delay event could have prolonged an activity, which may be demonstrated by revising the duration by the appropriate number of days.
4. Sometimes a delay event may be demonstrated by including the delay event itself as a new activity within the programme. If this method is chosen, the appropriate logic links must be introduced.

The resulting impacted as-planned programme will show the effect of the delay event on the critical path and, consequently, on the completion date. The revised completion date will, in turn, demonstrate the extension of time to which the claimant is entitled.

Whatever method of delay analysis is used, it is important to justify that is a suitable method. On many occasions, a claimant will adopt one method only for the respondent to state that a different method should be used. Sometimes there are good reasons for doing so, but frequently, this is just a reason for prolonging the making of a decision by the respondent. As with most matters when compiling a claim, it is best to deal with anticipated responses from a reviewer within the claim and effectively 'close the door' on them, rather than waiting for the response.

The Society of Construction Law recommends that the time impact method of delay analysis is used to demonstrate the effect of delay, so if you are going to use this method, reference to the Society of Construction Law's *Delay and Disruption Protocol* would provide sufficient justification. If it is preferred to use an alternative method, the reasons for doing so must be explained within the claim.

Chapter 7

Here is an example of how our example claim document could deal with the method of delay analysis:

SECTION 5
THE METHOD OF DELAY ANALYSIS

Introduction

This section describes the methodology used to demonstrate the effect of the delay event on the programme and, consequently, on the planned Time for Completion and thus demonstrates the extension of time to which the Contractor is entitled.

Basis of Delay Analysis

1. In order to demonstrate the effect of the delay on the Time for Completion, it is necessary to carry out a retrospective critical-path delay analysis. It is accepted that for the Contractor to be entitled to an extension of time, it must be shown that the Employer-caused delay event has affected the Time for Completion. Additionally, the Contractor is only entitled to claim for a time extension for Employer-caused delay events and, therefore, the Contractor has to ignore any other delays that are not as a result of such circumstances.

2. Considering the simple nature of the delay in question, an impacted as-planned delay analysis has been used to ascertain the effect of the delay. This is further explained as follows.

3. Certain activities on the programme are critical to the completion of the project within the contracted time. Such activities have no float, and any delay to these activities has a corresponding effect on the completion date. Such activities form the critical path.

4. If the start, for example, of an activity or sequence of activities on the critical path is delayed by the Employer by 20 days then, if the same logic and durations as the baseline programme are maintained, the completion date will be delayed by a corresponding 20 days. The Contractor would thus be entitled to an extension of time of 20 days under such circumstances.

5. If the Contractor causes additional delays, for example, by taking an additional 5-day duration to complete the activity, then the completion will be delayed by 25 days. The Contractor, however, would not be entitled to an extension of time for the 5-day delay of his own making, and his entitlement would remain at 20 days.

6. In an impacted as-planned programme, only the 20 days of delay brought about by the Employer would be impacted into the programme; the additional 5-day Contractor-caused delay would be ignored.
7. In addition, where there are many Employer-caused delays, not all of them will have a direct effect on the critical path and thus on the completion date. The effect of concurrency within these various Employer-caused delays is therefore taken into account by impacting all delay events and allowing the logic of the programme to determine the critical path and, consequently, the overall effect of the concurrent delays on the completion date.
8. Thus, by including only Employer-caused delays into an impacted as-planned programme, the Contractor will demonstrate his entitlement entirely as a result of the Employer-caused delay in that is the subject of this claim having made due allowance for any concurrency thereof. More importantly, such an analysis is unaffected by any delays attributable to the Contractor.
9. The impacted as-planned method is thus the chosen method of delay analysis used in this claim.

Baseline programme

1. As noted above, in order to complete an impacted as-planned analysis, there must be a reference programme or baseline against which the impact of delaying events can be measured.
2. The reference programme used for this analysis is the revised programme reference *RosTow BL-0 dated 1 March 2011*, which was approved by the Engineer on 9 March 2011[1] and is hereinafter referred to as the 'Un-Impacted Programme'.
3. The Un-Impacted Programme was updated following previous extension-of-time awards:
   a. Extension of Time No. 1 for Unforeseeable Physical Conditions which resulted in a substantial increase in the piling and foundation works and for which an extension of time to 21 January 2013 was awarded.[2]
   b. Extension of Time No. 2 for the late nomination of the facade contractor and for which an extension of time to 9 February 2013 was awarded[3]
4. The Un-Impacted Programme was approved by the Engineer on 5 March 2012[4] and is the programme used to demonstrate the effects of the delay claimed herein. This programme is hereinafter referred to as the 'EOT 2 Baseline Programme' and is included in this submission under Appendix B.

---

[1] Exhibit 4 – DWP letter reference ST/JW/Contr/036, dated 09/03/11
[2] Exhibit 6 – DWP letter reference ST/JW/Contr/126, dated 22/09/11
[3] Exhibit 7 – DWP letter reference ST/JW/Contr/295, dated 17/01/12
[4] Exhibit 8 – DWP letter reference ST/JW/Contr/345, dated 05/03/12

Chapter 7

Conclusion

1. An impacted as-planned method delay analysis has been chosen to demonstrate the effects of the delay on the Time for Completion. This is appropriate because of the relative simplicity of the delay event.
2. The delay event will be impacted into the latest approved as-planned programme, i.e. the EOT 2 Baseline Programme.
3. The method of impacting the various delay events into the impacted as-planned programme will be described in detail in Section 6 herein.

The method of delay analysis incorporating the above explanation could very well be dealt with in the section that provides details of the extension-of-time claim that follows, rather than in a separate section within the claim. Much here depends on the level of explanation necessary to justify the basis of the delay analysis. Exactly where this explanation is included in the claim document is not particularly important, but it is essential that a detailed explanation should be provided for the reviewer.

Checklist – The method of delay analysis

1. Introduction – the purpose of this Section
2. The method used to demonstrate the delay and the effect of the delay on the Time for Completion
3. Justification for the chosen method of delay analysis
4. The Baseline Programme
5. Conclusion – a summary of this Section

The Extension of Time Claim

Often the best way to deal with the event itself is to present the details through a chronology. The chronology should describe what happened and when it happened and provide substantiation of the events by way of reference to the project records. The project records should, of course, be included in an appendix for verification and reference. We should also split the chronology into sections that deal with cause and effect so that it is very clear that we have these essential elements covered. This, together with the substantiation provided by way of the documents included in the appendices, will ensure that we have complied with the CEES rule as discussed in detail in Chapter 5. The chronology and subsequent examination of entitlement related to the event should gradually lead the reviewer to the logical conclusion that the claimant is entitled to the time or amount claimed.

Chapter 7

Whatever method of delay analysis is chosen as being the most appropriate to demonstrate the effect, it is essential to include an explanation of the adopted method and exactly how the programme used to demonstrate entitlement was created. Such an explanation should be given in such a way that a non-expert planner can understand the methodology. It is often the case that the claim narrative is produced by one person and the delay-analysis programmes by another and it is also frequently the case that the two documents have little interrelationship or are sometimes even contradictory. On many occasions I have been presented with an entitlement programme and I have had absolutely no idea of the methodology used to create it, because the person writing the narrative has not taken the trouble to explain it. Possibly the logic would be obvious to an experienced planner, but it is quite possible that the reviewer of your claim document won't be an expert in this field and needs the benefit of an explanation. A step-by-step explanation of what has been done must be included at this point in the claim document and the resulting programme should be included in the appendices. The reviewer should certainly not have to guess the logic of, and the reasoning used in, the creation of the programme.

An important principle to bear in mind at this stage is that where two delays occur at the same time or concurrently, and one of the concurrent delays is the responsibility of the Contractor and the other is the responsibility of the Employer, the entitlement of the Contractor to additional payment for prolongation will probably be affected or negated altogether. The rationale of this principle is as follows:

1. If a delay is caused by the Employer, the Contractor is usually entitled to an extension of time and additional payment for the Contractor's prolongation costs, and the Employer is not entitled to deduct delay damages.
2. If a delay is caused by the Contractor, the Contractor is not entitled to an extension of time or additional payment for prolongation, and the Employer is entitled to deduct delay damages.
3. If there is a period of concurrent delay by both parties, the Contractor is entitled to an extension of time based upon the fact that had the Contractor not delayed the works, then the works would still have in any case been delayed due to the Employer-caused delay. This would give the Contractor relief from delay damages but would not provide entitlement to additional payment for prolongation costs.

As you might imagine, the above principles have been the subject of much debate, and there are many methods of demonstrating and disproving concurrent delays. In my experience, however, this basic principle is sufficient to use as a basis of entitlement for the submission and determination of the majority of claims and so, for the purposes of this book at least, let us consider them to be acceptable.

Chapter 7

An important investigation to make at this stage in the claim document, therefore, is whether the Contractor has caused concurrent
delays and if not, then the Contractor should demonstrate the fact. This
could be done by reference to the monthly reports or other records to
show that, at the time the delay was incurred, the Contractor was on
time. If the claimant is able to demonstrate this effectively, it will remove
the possibility of the respondent using concurrent delays as a reason
not to award additional payment for prolongation.

It is not unusual, when examining the events surrounding a claim,
and particularly the conditions of contract, to discover that the claimant
has not complied with the provisions of the contract in all respects.
Failure to submit notices or particulars within the specified time frames
is a fairly typical failing in this respect. In such cases, the claimant has
two choices: either ignore the failure and hope that the reviewer will not
notice the non-compliance, or acknowledge the issue and submit a
compelling argument as to why it should not affect his entitlement.
In the first case, the claimant might be lucky and the reviewer might not
realise that there is a weakness in the case, but it is not likely that a
more experienced reviewer will do so, should the matter escalate into a
dispute. Alternatively, and particularly if the failure was a condition
precedent to entitlement, the reviewer may reject the claim in its entirety
or may at best reduce the amount of the claim. This would require the
claimant to make a counter-response and to argue his case at a later
date which only serves to prolong the process. In the second case, the
claimant not only demonstrates his integrity by acknowledging the
failing but also presents his arguments at the outset, which should
serve to bring the matter to a quicker conclusion.

Let's have a look at how this could be dealt with in our example
claim.

SECTION 6
DETAILS OF THE CLAIM FOR AN EXTENSION
OF THE TIME FOR COMPLETION

Introduction

1. This Section examines the delay surrounding the alterations to the
   Electrical Transformer Room and subsequently the 'power-on date' that
   occurred during the construction period. It also sets out the cause of the
   delay and the effect on the planned sequence and timing of the Contractor's

activities by way of an impacted as-planned programme which demonstrates the effect on the Time for Completion.

2. The delay is demonstrated by reference to a chronology of events which are substantiated by the contemporaneous project records which are included herein under Appendix A.

3. The effect of the delay on the Contractor's intended programme has been demonstrated and a narrative included to explain exactly how the delay has been included in the impacted as-planned programme to demonstrate the effect on the overall Time for Completion.

4. Finally, the Contractor's entitlement to an award of an extension of the Time for Completion for this delay event is demonstrated by reference to the Contract.

The Cause

1. The Contract provides for the design to be provided by the Employer. The design includes an electrical transformer room on the ground floor in which the transformers are located. The Contractor constructed the Transformer Room in accordance with the revision-C drawings issued for construction on 5 July 2011[5] which are included for reference in Appendix C herein.

2. It should be noted that while the Employer is responsible for payment of the electricity authority's services and connection fees, Contract Specification No. 23-34-78 provides that the Contractor is responsible for coordinating the Works with the electricity authority.[6] The coordination included obtaining the electricity authority's acceptance of the Transformer Room prior to the installation of the transformers.

3. The Contractor wrote to the National Electricity Company on 2 January 2013 to inform them that the Electrical Transformer Room was complete and to request the National Electricity Company to inspect and approve the Transformer Room before 11 January 2013.[7]

4. On 8 January 2013, a National Electricity Company representative met with the Contractor and the Engineer's representative on site to inspect the Transformer Room and to discuss coordination details for the connection of the electricity supply to the transformers. During the visit, the National Electricity Company representative expressed concern that the Transformer Room did not appear to meet the National Electricity Company's requirements, but that he would need to verify this and inform the Contractor of his findings. The Site Progress Meeting minutes of 11 January 2013 recorded the following in relation to this matter: *NEC representative inspected GF Transformer Room on 8/1/13 and informed that he was concerned that it may not meet NEC specs. NEC rep to research and inform JCG.'*[8]

---

[5] Exhibit 5 – DWP Drawing Issue Transmittal No. 042, dated 05/07/11
[6] Exhibit 1 – Extract from Contract Specification No. 23-34-78
[7] Exhibit 10 – JCG letter reference P-1013/Let-1345/NEC, dated 02/01/13
[8] Exhibit 11 – Extract from minutes of Site Progress Meeting No. 51, dated 11/01/13

Chapter 7

5. On 14 January 2013 the National Electricity Company wrote to the Contractor and pointed out the following:
   a. The Electrical Transformer Room was of insufficient area.
   b. The external louvred doors which provide access to the Electrical Transformer Room were of the wrong dimensions[9].
6. On 15 January 2013 Johnson Construction Group wrote to the Dawson-Wilkinson Partnership enclosing the National Electricity Company's comments and closed the letter with the following statement: '*Obviously, these comments could have serious implications for the installation of the transformers and the subsequent energising of the Project. We therefore request that you instruct us as to how to proceed as soon as possible*'.[10]
7. The Contractor understands that several meetings were held between the Engineer, the Employer's designer and the National Electricity Company, during which the requirements of the National Electricity Company were resolved and finalised. The Contractor further understands that the problem occurred due to changes in the electricity authority's requirements which were introduced shortly after the design had been completed by the Employer's designer.
8. On 4 February 2013, the Engineer issued revised drawings for the area encompassing the Electrical Transformer Room and the adjacent Garbage Room and Maintenance Room with instructions to proceed with the revised works.[11] The drawings are included herein under Appendix F and the works detailed therein are summarised as follows:
   a. Demolition of the existing blockwork partition wall between the Electrical Transformer Room and the adjacent Garbage Room and reconstruction to provide 500 mm additional width to the Electrical Transformer Room.
   b. Demolition of the existing blockwork partition wall between the Electrical Transformer Room and the adjacent Maintenance Room and reconstruction to provide 250 mm additional width to the Electrical Transformer Room.
   c. Removal of the existing aluminium external louvred double doors, breaking out the existing concrete lintel, extending the door opening width by 350 mm, installing a new concrete lintel and new doors.
   d. Although not detailed on the drawings, this work also required making good to external granite cladding, which would be damaged by the formation of the extended external door opening.
9. Photographs included herein under Appendix D show that at the time the instruction was received, all works with the exception of second-fix electrical works and decoration were complete to the three rooms affected by the revisions.

---

[9] Exhibit 12 – NEC letter reference 13/0978NC/3945, dated 14/01/13
[10] Exhibit 13 – JCG letter reference P-1013/Let-1362, dated 15/01/13
[11] Exhibit 18 – DWP Drawing Issue Transmittal No. 139, dated 04/02/13

The Effect

1. The Daily Site Report of 5 February 2013 records that the Contractor commenced demolition of the blockwork partition walls on this date.[12]
2. It should be noted that the manufacture of the aluminium louvred external doors was a long-lead item by the Contractor's aluminium subcontractor who required to take site measurements of the opening before manufacture could commence. The Contractor took steps to mitigate the lead time by fabricating a steel template around which the blockwork was rebuilt following the breaking out of the enlarged opening. This methodology provided the aluminium subcontractor with fixed dimensions from which the doors could be manufactured, which allowed fabrication to take place in parallel with the alteration works.
3. As recorded in the Daily Site Reports between 5 and 13 February 2013, the Contractor also mitigated the effects of this delay by working extended shifts and weekends until the essential work required to enable the electricity authority to install the transformers was completed.[13]
4. As recorded in the Daily Site Report for 13 February 2013, the Contractor completed all the work comprising demolition, blockwork, plasterwork, concrete-plinth alterations and the installation of the external louvred doors on 13 February 2013, leaving only the decorations and making good to the external granite cladding outstanding.[14]
5. The Contractor arranged for the National Electricity Company to inspect the alterations on 14 February 2013[15] and, following the inspection, the National Electricity Company confirmed that the Electrical Transformer Room was acceptable and that the Contractor could proceed with the installation of the transformers. This was confirmed retrospectively in the National Electricity Company's letter dated 16 February 2013.[16]
6. The Daily Site Reports for 15 and 25 February 2013 record that the Contractor subsequently started to install and connect the transformers on 15 February 2013[17] and completed the operation on 25 February 2013.[18] The National Electricity Company connected the mains supply and energised the transformers on 26 February 2013.[19]
7. It should be noted that, as the project is located on a 'brown field' area, which is being developed as part of a larger programme in parallel with new local infrastructure, no mains electricity was available to the Contractor through the construction period, and his temporary site electrical power was provided by generators. Testing and commissioning

---

[12] Exhibit 19 – JCG Daily Site Reports, dated 05/02/13 to 13/02/13
[13] Exhibit 19 – JCG Daily Site Reports, dated 05/02/13 to 13/02/13
[14] Exhibit 19 – JCG Daily Site Reports, dated 05/02/13 to 13/02/13
[15] Exhibit 21 – JCG Daily Site Report, dated 14/02/13
[16] Exhibit 22 – NEC letter reference 13/0978NC/3973, dated 16/02/13
[17] Exhibit 23 – JCG Daily Site Report, dated 15/02/13
[18] Exhibit 24 – JCG Daily Site Report, dated 25/02/13
[19] Exhibit 25 – JCG Daily Site Report, dated 26/02/13

Chapter 7

of the building are not possible by the use of power from generators and, consequently, the connection of the permanent electricity supply and the energising of the transformers to provide 'clean' power were necessary to enable the testing and commissioning to start. Thus, following the connection of the permanent power on 26 February 2013, the Contractor was able to start the testing and commissioning operations on 27 February 2013.[20]

8. The effect of the incorrect design of the Electrical Transformer Room was to delay the start of testing and commissioning operations from the planned date of 24 January 2013 to 27 February 2013, a delay of 34 days.

9. The Daily Site Report for 10 March 2013 records that the Contractor finished testing and commissioning on 10 March 2013, a duration of 12 days.[21] This is a 4-day reduction from the originally planned duration of 16 days and demonstrates that the Contractor mitigated the delay from 35 days to 31 days.

10. The following section demonstrates how the delays described above have been included in the impacted as-planned programme, in order to demonstrate the effect on the critical path and thus on the overall Time for Completion.

Delay Analysis

1. The current programme at the time of this delay event has been used to impact this delay event is the EOT 2 Baseline Programme, a copy of which is included herein under Appendix B.

2. Extracts from the Monthly Report for December 2012 included herein under Appendix E include the Contractor's as-built programme and a comparison of this to the EOT 2 Baseline Programme. The Monthly Report demonstrates that the Contractor was generally on target with no delays to the critical-path activities at the time when the alterations to the Electrical Transformer Room became apparent during early to mid January 2013. No concurrent Contractor-caused delays can therefore be considered to affect the impacted as-planned programme or the Contractor's entitlement to an extension of time.

3. The EOT 2 Baseline Programme has been impacted to produce an impacted as-planned programme which is included herein under Appendix G and which demonstrates the effect of the delay event on the Time for Completion.

---

[20] Exhibit 26 – JCG Daily Site Report, dated 27/02/13
[21] Exhibit 27 – JCG Daily Site Report, dated 10/03/13

4. The impacted as-planned programme has been created from the baseline programme as follows:
   a. Introduction of a new activity: 'Resolve NEC requirements and issue drawings'.
      i. Start – 14 January 2013, the date the National Electricity Board notified the Contractor of the requirements.[22]
      ii. Finish – 4 February 2013, the date the Engineer issued the revised drawing[23]
   b. Introduction of a new activity: 'Alterations to Elec. Trans. Room'.
      i. Start – 5 February 2013, a successor to the drawing issue of 4 February 2013.
      ii. Finish – 13 February 2013, the actual finish date.[24]
   c. The single-day activity 'NEC inspect Elec. Trans. Room' moved from 10 January 2013 to the actual date of 14 February 2013.[25]
   d. The activity 'Install and commission transformers'.
      i. Start – changed to the actual date of 15 February 2013.[26]
      ii. Finish – changed to the actual date of 25 February 2013[27]
   e. The milestone 'Power On'.
      i. Changed to the actual date of 26 February 2013.[28]
   f. The activity 'Test and Commission'.
      i. Start date changed to the actual date of 27 February 2013.[29]
      ii. Finish date changed to the actual date of 10 March 2013, reflecting the reduced duration from 16 days as planned to 12 days actual[30]
5. The impacted as-planned programme, produced as a result of the above modifications to the baseline programme, shows a revised completion date of 11 March 2013, the day following the completion of the testing and commissioning.
6. The Contractor therefore claims an extension of the Time for Completion of 30 days from 9 February 2013 to 11 March 2013 for the delay event included in this claim.

Entitlement to Extension of Time

1. The Contractor's entitlement to an extension of the Time for Completion is contained in the following clauses, which are reproduced in full in Section 4 herein.

---

[22] Exhibit 12 – NEC letter reference 13/0978NC/3945, dated 14/01/13
[23] Exhibit 18 – DWP Drawing Issue Transmittal No. 139, dated 04/02/13
[24] Exhibit 19 – JCG Daily Site Reports, dated 05/02/13 to 13/02/13
[25] Exhibit 21 – JCG Daily Site Report, dated 14/02/13
[26] Exhibit 23 – JCG Daily Site Report, dated 15/02/13
[27] Exhibit 24 – JCG Daily Site Report, dated 25/02/13
[28] Exhibit 25 – JCG Daily Site Report, dated 26/02/13
[29] Exhibit 26 – JCG Daily Site Report, dated 27/02/13
[30] Exhibit 27 – JCG Daily Site Report, dated 10/03/13

2. Sub-Clause 1.9 [*Delayed Drawings or Instructions*] provides that

'*The Contractor shall give notice to the Engineer whenever the Works are likely to be delayed or disrupted if any necessary drawing or instruction is not issued to the Contractor within a particular time, which shall be reasonable. The notice shall include details of the necessary drawing or instruction, details of why and by when it should be issued, and details of the nature and amount of the delay or disruption likely to be suffered if it is late.*'

3. Johnson Construction Group initially wrote to the Dawson-Wilkinson Partnership on 15 January 2013 enclosing the National Electricity Company's comments on the Electrical Transformer Room and requested that the Dawson-Wilkinson Partnership '*instruct us as to how to proceed as soon as possible*'[31].

4. In the subsequent weeks Johnson Construction Group wrote to the Dawson-Wilkinson Partnership on several occasions regarding this subject and also drew the Dawson-Wilkinson Partnership's attention to the issue in the site progress meetings as follows:

   a. Letter dated 22 January 2013 from Johnson Construction Group to the Dawson-Wilkinson Partnership confirming that the installation of the transformers was currently on hold and that this will cause the testing and commissioning and handover to be delayed and requesting instructions from the Dawson-Wilkinson Partnership[32].

   b. The minutes of the Site Progress Meeting of 25 January 2013 state '*JCG requested advice as to the situation with the ET Room. DWP advised that Corrie were in discussions with NEC and it is possible that the ET Room will need to be altered*'[33].

   c. Letter dated 31 January 2013 from Johnson Construction Group to the Dawson-Wilkinson Partnership stating that '*We confirm that the continued delay in receiving instructions with regard to the Electrical Transformer Room will cause a delay to the handover of the project*'[34].

5. Sub-Clause 1.9 [*Delayed Drawings or Instructions*] continues as follows:

'*If the Contractor suffers delay and/or incurs Cost as a result of a failure of the Engineer to issue the notified drawing or instruction within a time which is reasonable and is specified in the notice with supporting details, the Contractor shall give a further notice to the Engineer and shall be entitled subject to Sub-Clause 20.1* [Contractor's Claims] *to: ... (c) an extension of time for any such delay, if completion is or will be delayed, under Sub-Clause 8.4* [Extension of Time for Completion] *...*'.

---

[31] Exhibit 13 – JCG letter reference P-1013/Let-1362, dated 15/01/13
[32] Exhibit 14 – JCG letter reference P-1013/Let-1367, dated 22/01/13
[33] Exhibit 15 – Extract from minutes of Site Progress Meeting No. 52, dated 25/01/13
[34] Exhibit 16 – JCG letter reference P-1013/Let-1373, dated 31/01/13

The contemporaneous programme showed that the 'power-on' milestone was due on 23 January 2013. Therefore, in order not to delay this date, the Engineer should have issued an instruction in such time as to allow for the alteration works necessary to satisfy the requirements of the National Electricity Company to be completed and the transformers to be installed in time to meet this date. As detailed earlier in this section, the revised drawings were issued on 4 February 2013 and the actual power-on date was 26 February 2013, which equates to a duration of 23 days. The following calculation shows the date that the instructions should have been received by the Contractor in order to meet the original power-on date:

EOT 2 programmed 'power-on date':	23 January 2013
LESS: Install and commission transformers (Actual dates: 15/02/13–25/02/13)	11-day duration
	11 January 2013
LESS: NEC inspection	1-day duration
	10 January 2013
LESS: Demolition and alteration work (Actual dates: 05/02/13–13/02/13)	9-day duration
Latest date for issue of revised drawings:	1 January 2013

6. The actual date of the issue of the revised drawings was 4 February 2013 some 34 days later than the date after which a delay would occur to the programme.
7. The Contractor is therefore entitled to an extension to the Time for Completion under the provisions of Sub-Clause 1.9 [*Delayed Drawings or Instructions*].
8. Sub-Clause 4.7 [*Setting Out*] provides that

   '*The Contractor shall set out the Works in relation to original points, lines and levels of reference specified in the Contract or notified by the Engineer. The Contractor shall be responsible for the correct positioning of all parts of the Works, and shall rectify any error in the position, levels, dimensions or alignment of the Works.*'

   The Contractor, having constructed the Electrical Transformer Room to the revision-C drawings issued for construction on 5 July 2011 (which are included for reference in Appendix C herein) had complied with this provision[35].
9. Sub-Clause 4.7 [*Setting Out*] goes on to state that

   '*The Employer shall be responsible for any errors in these specified or notified items of reference, but the Contractor shall use reasonable efforts to*

---

[35] Exhibit 5 – DWP Drawing Issue Transmittal No. 042, dated 05/07/11

verify their accuracy before they are used. If the Contractor suffers delay and/or incurs Cost from executing work which was necessitated by an error in these items of reference, and an experienced contractor could not reasonably have discovered such error and avoided this delay and/or Cost, the Contractor shall give notice to the Engineer and shall be entitled subject to Sub-Clause 20.1 [Contractor's Claims] to: … (c) an extension of time for any such delay, if completion is or will be delayed, under Sub-Clause 8.4 [Extension of Time for Completion]'.

10. The Contractor, having constructed the Electrical Transformer Room to the revision-C drawings, complied with the requirements with regard to 'the correct positioning of all parts of the Works' in accordance with these drawings. The Contractor carries no design liability under the Contract and while the Contractor acknowledges that he is responsible for coordination with the electricity authority, submits that such coordination is limited to liaising with the electricity authority in terms of the timing of the authority's work and access to the Site, but does not extend to verifying the correctness of the Employer's design. The Contractor is therefore entitled to an extension to the Time for Completion under the provisions of Sub-Clause 4.7 [*Setting Out*].

11. Sub-Clause 8.5 [*Delays Caused by Authorities*] provides that

'If the following conditions apply, namely: (a) the Contractor has diligently followed the procedures laid down by the relevant legally constituted public authorities in the Country, (b) these authorities delay or disrupt the Contractor's work, and (c) the delay or disruption was Unforeseeable, then this delay or disruption will be considered as a cause of delay under sub-paragraph (b) of Sub-Clause 8.4 [Extension of Time for Completion].'

12. The Contractor was not privy to the meetings between the Engineer, the Employer's designer and the National Electricity Company and is therefore unaware of the reasons why the design contained in the issued-for-construction drawings, to which the Contractor con-structed the Electrical Transformer Room, did not comply with the requirements of the electricity authority. The Contractor submits that the authority's procedures were diligently followed insofar as these relate to the Contractor's responsibility to coordinate the Works. Should it be the case that the National Electricity Company's insistence on the alterations to the design was without just cause, then this should be regarded as delay and disruption by the electricity authority, and the Contractor is consequently entitled to an exten-sion to the Time for Completion under the provisions of Sub-Clause 8.5 [*Delays Caused by Authorities*]. The Engineer was party to the discussions and meetings between the designer and the electricity authority, so the Contractor must leave the determination of this issue to the Engineer.

Chapter 7

13. Sub-Clause 8.4 [*Extension of Time for Completion*] provides that

   '*The Contractor shall be entitled subject to Sub-Clause 20.1* [Contractor's Claims] *to an extension of the Time for Completion if and to the extent that completion for the purposes of Sub-Clause 10.1* [Taking Over of the Works and Sections] *is or will be delayed by any of the following causes: (a) a Variation (unless an adjustment to the Time for Completion has been agreed under Sub-Clause 13.3* [Variation Procedure]*) … (b) a cause of delay giving an entitlement to extension of time under a Sub-Clause of these Conditions … (e) any delay, impediment or prevention caused by or attributable to the Employer, the Employer's Personnel, or the Employer's other contractors on the Site*'.

14. The issue of the revised drawings for the Electrical Transformer Room constitutes a Variation to the Contract. The delay in issuing instructions relating to the alterations to the Electrical Transformer Room constitutes a cause of delay '*giving an entitlement to extension of time*' under Sub-Clauses 1.9 [*Delayed Drawings or Instructions*] and 4.7 [*Setting Out*] of the Conditions. The delays in relation to the alterations to the Electrical Transformer Room can also be regarded as being '*delay, impediment or prevention caused by or attributable to the Employer's Personnel or the Employer's other contractors on the Site*', in the form of the Employer's designers and the electricity company respectively. The Contractor is therefore entitled to an extension to the Time for Completion under the provisions of Sub-Clause 8.4 [*Extension of Time for Completion*].

15. Sub-Clause 8.4 [*Extension of Time for Completion*] continues as follows:

   '*If the Contractor considers himself to be entitled to an extension of the Time for Completion, the Contractor shall give notice to the Engineer in accordance with Sub-Clause 20.1* [Contractor's Claims]*.*'

16. Sub-Clause 20.1 [*Contractor's Claims*] provides that:

   '*If the Contractor considers himself to be entitled to any extension of the Time for Completion … in connection with the Contract, the Contractor shall give notice to the Engineer, describing the event or circumstance giving rise to the claim. The notice shall be given as soon as practicable, and not later than 28 days after the Contractor became aware, or should have become aware, of the event or circumstance.*'

17. Johnson Construction Group initially wrote to the Dawson-Wilkinson Partnership on 15 January 2013 enclosing the National Electricity Company's comments on the Electrical Transformer Room and requested that '*you instruct us as to how to proceed as soon as possible.*'[36]

---

[36] Exhibit 13 – JCG letter reference P-1013/Let-1362, dated 15/01/13

18. Johnson Construction Group also wrote to the Dawson-Wilkinson Partnership on 22 January 2013 confirming that the installation of the transformers was currently on hold and that this would cause the testing and commissioning and handover to be delayed.[37]

19. The minutes of the Site Progress Meeting of 25 January 2013 state: '*JCG requested advice as to the situation with the ET Room. DWP advised that Corrie were in discussions with NEC and it is possible that the ET Room will need to be altered.*[38]

20. Johnson Construction Group wrote to the Dawson-Wilkinson Partnership on 31 January 2013 stating that '*We confirm that the continued delay in receiving instructions with regard to the Electrical Transformer Room will cause a delay to the handover of the project.*'[39]

21. The above confirms that the Engineer was fully aware of both the circumstances and the probable delay which would be caused by this issue.

22. Sub-Clause 20.1 [*Contractor's Claims*] continues as follows:

'*If the Contractor fails to give notice of a claim within such period of 28 days, the Time for Completion shall not be extended ... and the Employer shall be discharged from all liability in connection with the claim.*'

23. The Contractor gave notice that he considered himself to be entitled to an extension of the Time for Completion by way of his letter to the Engineer on 17 February 2013 which states:

'*We confirm receipt of the revised drawings for the Electrical Transformer Room area issued under Transmittal No. 139 dated 4 February 2013 and give notice that the delays in resolving the situation surrounding the Electrical Transformer Room to the satisfaction of the National Electricity Company will delay the completion date of the project.*

We consider that this constitutes a claimable event for an extension of time and prolongation costs under Clause 20.1 and will submit our claim in due course.'[40]

24. It could be argued retrospectively that the Contractor should have become aware that the problems surrounding the Electrical Transformer Room would delay the Time for Completion on 14 January 2013, this being the date when the National Electricity Company wrote to detail the items that were not in accordance with the National Electricity Company's requirements.[41] At this point, however, it was not actually apparent to the Contractor that a delay would in fact occur. This is confirmed by reference to the Contractor's letter of 15 January 2013

[37] Exhibit 14 – JCG letter reference P-1013/Let-1367, dated 22/01/13
[38] Exhibit 15 – Extract from minutes of Site Progress Meeting No. 52, dated 25/01/13
[39] Exhibit 16 – JCG letter reference P-1013/Let-1373, dated 31/01/13
[40] Exhibit 20 – JCG letter reference P-1013/Let-1421, dated 17/02/13
[41] Exhibit 12 – NEC letter reference 13/0978NC/3945, dated 14/01/13

which states that: '*Obviously, these comments could have serious implications for the installation of the transformers and the subsequent energising of the Project. We therefore request that you instruct us as to how to proceed as soon as possible.*' [42]

25. The Contractor was not party to the subsequent meetings held between the Engineer, the Employer's designers and the National Electricity Company and, apart from advice received from the Engineer and recorded in the Progress Meetings that a solution was being developed between the Employer's designers and the National Electricity Company, the Contractor was unaware of what the solution would be, or of the effects of such a solution on the programme.

26. It could also be argued that, when the programmed date of 11 January 2013 for the start of the installation of the transformers had passed, the Contractor should have provided notice that he considered himself entitled to an extension to the Time for Completion because of the delay.

27. The Contractor acknowledges that the formal notice of 17 February 2013 was not submitted within the 28-day period, required by the Contract, from receipt of the advice from the National Electricity Board that the Transformer Room did not comply with their requirements, but submits that the events were recorded in several letters and minutes of the Progress Meetings, so the Engineer was fully aware of the circumstances and the possible effects. The Contractor also submits that, due to the Engineer's discussions with the Employer's designers and the National Electricity Company, the Engineer was, at this time, more aware than the Contractor of the solution to the problem and thus the probability of delay.

28. The Contractor submits that the requirement to provide notice to the Engineer within the 28-day period stipulated in the Contract is included to enable the Engineer to investigate the circumstances and consider solutions to mitigate the delay. In this instance, as the Engineer was made aware of the circumstances by the Contractor before the notice was given and as the Engineer was responsible for issuing instructions to the Contractor to resolve the delay, the delay in submitting a formal notice under the provisions of Sub-Clause 20.1 [*Contractor's Claims*] should not, fairly, reasonably and with due regard to the particular circumstances, be regarded by the Engineer as grounds to deny the Contractor's entitlement to an extension of the Time for Completion.

29. Sub-Clause *8.7* [*Delay Damages*] provides that

'*If the Contractor fails to comply with Sub-Clause 8.2* [Time for Completion]*, the Contractor shall subject to Sub-Clause 2.5* [*Employer's Claims*] *pay delay damages to the Employer for this default.*'

Chapter 7

---

[42] Exhibit 13 – JCG letter reference P-1013/Let-1362, dated 15/01/13

The Contractor submits that because he has an entitlement to an extension of the Time for Completion for the delay event described in this claim, he is relieved of the obligation to pay delay damages.

30. Sub-Clause *3.5 [Determinations]* provides that

'*Whenever these Conditions provide that the Engineer shall proceed in accordance with this Sub-Clause 3.5 to agree or determine any matter … If agreement is not achieved, the Engineer shall make a fair determination in accordance with the Contract, taking due regard of all relevant circumstances*'.

This claim is therefore submitted to the Engineer in order that agreement may be reached, or a fair determination of the extension of the Time for Completion to which the Contractor is entitled may be made.

Conclusion

1. The requirements of the National Electricity Company resulted in the Engineer being obliged to instruct the Contractor to revise the already-constructed Electrical Transformer Room to meet these requirements.

2. The instruction was issued to the Contractor at a time when the instructed revised works affected the programmed dates for installation and commissioning of the electrical transformers, the subsequent power-on date and the start of the testing and commissioning activities.

3. The Contractor took various steps to minimise the time taken to carry out the alteration work and also mitigated the delay by completing the testing and commissioning activities within a reduced duration from that which was planned.

4. The Contractor has demonstrated the effect of the delay by producing a delay analysis, which demonstrates that the delay caused to the completion of the Project for this event is 30 days, causing a revised completion date of 11 March 2013.

5. The Contractor has demonstrated that there were no Contractor-caused delays concurrent with this delay event.

6. The Contractor's entitlement to an extension of the Time for Completion is contained in the following clauses:
   a. Sub-Clause 1.9 [*Delayed Drawings or Instructions*]
   b. Sub-Clause 4.7 [*Setting Out*]
   c. Sub-Clause 8.4 [*Extension of Time for Completion*]
   d. Sub-Clause 8.5 [*Delays Caused by Authorities*]

7. The Contractor has complied with the provisions of Sub-Clause 20.1 [*Contractor's Claims*] to a fair and reasonable extent and submits that in the particular circumstances, any failure to submit a formal notice within the time frame provided in the Contract cannot, fairly and reasonably, be regarded as grounds to deny the Contractor's entitlement to an extension of the Time for Completion.

8. The reasons entitling the Contractor to an extension of the Time for Completion for the delay has negated the Employer's entitlement to the payment of delay damages.
9. The Contractor therefore claims an extension of the Time for Completion to 11 March 2013 for the delay event included in this claim and requests the Engineer to make a fair determination in accordance with the Contract for this event.

Checklist – Claim for an extension to the Time for Completion

1. Introduction – the purpose and an outline of this Section
2. The Cause
3. The Effect
4. Delay analysis
5. Explanation of how the delay analysis has been created
6. Revised Time for Completion
7. Concurrent delays
8. Entitlement to an extension of time under the Contract
9. Demonstration that the claimant has complied with conditions precedent
10. Reasoned arguments in cases where claimant has not complied with conditions precedent
11. Negation of the Employer's entitlement to delay damages
12. Reminder of the Engineer's obligations
13. Conclusion – a brief summary of the section
14. Substantiation by reference to the project records

Chapter 7

Chapter 8

The Claim for Additional Payment

It is common practice for contractors to link claims for additional payment for prolongation costs to claims for extensions of time and to present both as one single claim. While this may be appropriate in straightforward circumstances, consideration should be given to dealing with the two subjects separately, on the basis that the award of the extension of time is invariably easier to agree and determine because, at this stage at least, it does not cost the Employer money. Thus, a claim for an extension of time alone may be dealt with relatively quickly as opposed to protracting the matter while details of monetary calculations for prolongation are considered, discussed and negotiated. It is common practice for different people to review claims for time and payment, so the submission of separate claims can also help the reviewing party to organise their work so that the time and cost claims are reviewed in parallel. Additionally, while the Contractor may claim a certain number of days for an extension of time, it could well be that the Engineer will determine that less time is warranted and make an appropriate award. Thus, if the contractor's prolongation claim is linked to the extended period as it inevitably is, then the calculations of the additional payment will have to be revised and resubmitted in accordance with the Engineer's determination. If it is decided to submit separate time and monetary claims, care must be taken to comply with the time frames prescribed in the Contract for the submission of the claim for prolongation costs. A sensible way of dealing with this is to submit the two claims at the same time, but as separate and discrete submissions. For the purposes of this book and in order to illustrate a typical claim for prolongation costs, we shall not, however, take the sensible course of action and we will include a claim for additional payment within our example claim.

As with all claims, it is necessary to establish cause, effect and entitlement and back these up with substantiation (CEES) in order to justify both the claimant's entitlement and the quantum of the claim. Having already completed the extension-of-time claim in our example,

Construction Claims & Responses: Effective Writing & Presentation, Second Edition. Andy Hewitt.
© 2016 John Wiley & Sons, Ltd. Published 2016 by John Wiley & Sons, Ltd.

much of the work associated with establishing cause and effect has already been completed and it remains to deal with entitlement to additional payment (which may be different from entitlement to an extension of time) and the quantum of the claim.

An important point to bear in mind is that, while the claimant is obliged to demonstrate his case on the balance of probabilities, this does not mean that a claim for additional payment may be demonstrated by theoretical or notional calculations involving estimates or assessments. In other words, the claimant has to prove his case, based upon fact, and this, in turn, means that quantities, durations, rates and prices must all be demonstrable and substantiated.

In our example, the claim for additional payment will be based upon the fact that the Contractor has suffered damage due to the extended time that he has been obliged to remain on site. The conclusions of the cause and effect from the previous claim section which deals with the extension of time should therefore be repeated in the section that deals with additional payment. In this case, however, the effect will need to be related to the financial damage incurred rather than time. Generally, the claim for additional payment in the case of prolongation may be categorised as follows:

1. Site-establishment costs including site staff, site establishment, transport, plant, equipment and the necessary running and maintenance of such items (a detailed list of such items is given in the form of a checklist later in this chapter)
2. Contractual costs such as insurances and performance guarantees
3. Head-office overheads and profit
4. Finance costs

Explanations should be given as to the nature of the additional costs incurred and the costs applicable to the claim should be demonstrated in the calculations.

Prolongation costs are mainly related to time, so it is important to demonstrate the number of days on which the calculations are based – a simple calculation included within the narrative will usually suffice for this. It is a good idea to ensure that all cost calculations are based upon a cost per day so that if, for example, a reduced extension of time is agreed or determined, it is an easy task to recalculate the claim at a later date by a simple revision of the number of days upon which the claim is based. As with most narratives, it is important to describe the logic and methodology used in the financial calculations in a manner that will enable the reviewer to understand the process and lead him to a logical conclusion.

A fairly basic point that is sometimes misunderstood is that while the Contractor's costs should be for the extension of time period, the costs were actually incurred at the time of the delay and not during the extended contract period. This may be better explained as follows:

Period of delay – 1 January to 30 January	31 days
Completion date before extension of time	1 June
Completion date after delay analysis	21 June
Extended Time for Completion (1 June to 21 June)	20 days

The prolongation costs should therefore be for those costs incurred for 20 days during January and not for those incurred during June.

An important point worth mentioning at this juncture is that different contracts provide for different methods of financial recovery. Where a contract provides for 'loss and/or expense' to be recovered, this enables both the Contractor's costs and his losses to be taken into account. Some contracts do not allow for the claimant to recover financing charges and some allow for recovery of profit. As always, the place to check for the basis of the financial claim is the contract in question. When a contract refers to 'cost', this should usually be *actual costs incurred* by the Contractor and not, as is sometimes assumed, the estimated costs derived from the General Items or Preliminaries sections of the bills of quantities. FIDIC includes the following definition under Sub-Clause 1.1.4.3:

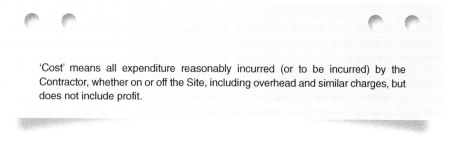

'Cost' means all expenditure reasonably incurred (or to be incurred) by the Contractor, whether on or off the Site, including overhead and similar charges, but does not include profit.

Thus, in such a situation and in order for a claim to succeed, the claimant must demonstrate his costs by reference to the resources deployed and invoices or the like to substantiate the *actual costs* of prolongation.

Detailed calculations, usually in the form of spreadsheets, should be appended, in order to demonstrate how the additional payment amount has been calculated. Such spreadsheets should be consistent with the narrative in terms of titles, column headings and the like and, if explanations are necessary, explanatory notes should be added, either to the individual spreadsheets, to the narrative or to both.

There are several ways of calculating head-office overheads and profit that have become established and recognised within the

Chapter 8

industry, and these include the Hudson formula and the Emden formula. Both provide a method of calculation based on the claimant's previously achieved levels of overheads and profit, information on which may usually be obtained from the Contractor's audited accounts. An example of Emden's formula is shown in the example claim later in this chapter.

Substantiation of the claimed costs should be included by way of invoices or the like. Such substantiation should be included in a separate appendix, and the invoices should be individually referenced and cross-referenced back to the calculations, so that the reviewer may easily verify the costs. Invoices for time-related costs will generally be for monthly amounts so, because we need to establish a daily cost, the calculation of monthly costs to daily costs must also be demonstrated. This may be done either by a spreadsheet calculation, or by annotating the invoice with a quick handwritten calculation similar to that shown in the following example:

Monthly invoice total	£ 1450
Days in invoiced month:	31 days
Cost per calendar day:	£ 46.77 per day

We have discussed previously how a claim should comprise a stand-alone document so that a reviewer has all the information required to verify the matters and to reach a conclusion. Sometimes, however, a claim for prolongation costs on a large project over a long period will involve thousands of individual costs items. In such a case, the inclusion of documentation to substantiate each individual cost would both be onerous and could lead to voluminous levels of documents being included within the submission. In these days of computerised accounting systems it is usual for a company to be able to produce monthly cost summaries for items specifically identified as being claimable. It could therefore be appropriate to use summaries from the project accounts within the claim and to offer the Engineer the opportunity to audit the costs, usually at the place where the accounts are maintained. In such a case, a reasonable Engineer should be prepared to carry out an audit, rather than checking each separate line item. Of course, if the audit throws up a large number of errors or the costs cannot be verified it may be necessary to go back to first principles.

It should be noted that if extensions of time are calculated in calendar days as opposed to working days, as they usually are, then the costs must be reconciled to the same basis of calendar days. Some parties consider that costs per working day provide a more realistic approach and, if this is the case, then care must be taken to ascertain the number of working days out of the extended period and to use this number as the multiplier of the daily cost.

The narrative should include explanations as to the basis of any calculations that are not absolutely self-explanatory or are not clearly demonstrated on the spreadsheets. Note how the example narrative contains several small, clearly explained calculations to demonstrate how certain figures have been ascertained.

Finally, a summary of the claimed amounts should be included in the narrative and a conclusion made in the form of a summary of the foregoing. Let's see how these principles could be included in our example claim.

SECTION 7
DETAILS OF THE CLAIM FOR
ADDITIONAL PAYMENT

Introduction

1. The cause of the Contractor's claim for additional payment is the delay event demonstrated in Section 6 herein, which obliged the Contractor to continue to maintain resources on site and contribute to head-office running costs for a period greater than was originally intended.

2. The financial effect of this event was that the Contractor incurred additional costs in maintaining his site establishment, providing finance for the works, providing insurances and the like and providing a contribution to his head-office overheads. The Contractor was also prevented from earning a contribution from other projects through having his resources tied up on the project for the extended period.

3. This Section comprises the Contractor's claim for additional payment for the prolongation costs incurred as a result of the delay and the consequent extension of time.

4. It should be noted that the Contractor's claim for the demolition and alteration works to the Electrical Transformer Room has been submitted under a separate variation claim. The variation claim includes only for the measured works and does not include any prolongation costs, which are claimed entirely herein.

Chapter 8

The Contractor's Entitlement to Additional Payment

1. The Contract provisions that provide entitlement to additional payment are as follows.

    Sub-Clause 1.9 [*Delayed Drawings or Instructions*] provides that '*If the Contractor … incurs Cost as a result of a failure of the Engineer to issue the notified drawing or instruction within a time which is reasonable and is specified in the notice with supporting details, the Contractor shall give a further notice to the Engineer and shall be entitled subject to Sub-Clause 20.1* [Contractor's Claims] *to:… (b) payment of any such Cost plus reasonable profit, which shall be included in the Contract Price.*'

2. Section 6 herein has examined the subject of delayed drawings and instructions in detail and has substantiated that the Engineer failed to issue the drawings and instructions for the work that was required to make the Electrical Transformer Room acceptable to the electricity authority in such time to enable the Contractor to comply with the previously established Time for Completion. The Contractor is therefore entitled to additional payment under the provisions of Sub-Clause 1.9 [*Delayed Drawings or Instructions*].

3. Sub-Clause 4.7 [*Setting Out*] provides that

    '*If the Contractor … incurs Cost from executing work which was necessitated by an error in these items of reference, and an experienced contractor could not reasonably have discovered such error and avoided this … Cost, the Contractor shall give notice to the Engineer and shall be entitled subject to Sub-Clause 20.1* [Contractor's Claims] *to … (b) payment of any such Cost plus reasonable profit, which shall be included in the Contract Price.*'

4. Section 6 herein has examined the issue of setting out in detail and has substantiated that the Contractor set out and constructed the Electrical Transformer Room in accordance with the issued-for-construction drawings provided by the Engineer. The Contractor is therefore entitled to additional payment under the provisions of Sub-Clause 4.7 [*Setting Out*].

5. Sub-Clause 20.1 [*Contractor's Claims*] provides that

    '*If the Contractor considers himself to be entitled to any additional payment, under any Clause of these Conditions or otherwise in connection with the Contract, the Contractor shall give notice to the Engineer, describing the event or circumstance giving rise to the claim. The notice shall be given as soon as practicable, and not later than 28 days after the Contractor became aware, or should have become aware, of the event or circumstance.*'

6. Section 6 herein provides full details and substantiation of the events and correspondence issued by the Contractor to the Engineer in connection with the Electrical Transformer Room.
7. Sub-Clause 20.1 [*Contractor's Claims*] continues as follows:

   'If the Contractor fails to give notice of a claim within such period of 28 days, the Time for Completion shall not be extended … and the Employer shall be discharged from all liability in connection with the claim.'

8. Section 6 herein provides full details and substantiation of the events and notices provided by the Contractor in this respect and acknowledges that the Contractor did not provide formal notice that he considered himself entitled to additional payment within the 28-day period. The Contractor submits that, for the same reasons included in Section 6 herein, the delay in submitting a formal notice under the provisions of Sub-Clause 20.1 [*Contractor's Claims*] should not, fairly and reasonably, be regarded by the Engineer as cause to deny the Contractor's entitlement to additional payment.
9. Sub-Clause 3.5 [*Determinations*] provides that

   'Whenever these Conditions provide that the Engineer shall proceed in accordance with this Sub-Clause 3.5 to agree or determine any matter … If agreement is not achieved, the Engineer shall make a fair determination in accordance with the Contract, taking due regard of all relevant circumstances.'

   This claim is therefore submitted to the Engineer in order that agreement or a fair determination of the additional payment to which the Contractor is entitled may be reached.
10. In addition to the contractual provisions, the principles of recovery, where one party to a contract has defaulted, are well established in law. Essentially, the aggrieved party is entitled by an award of money to be put back in the position in which it would have been had the contract been performed as originally envisaged. This would normally comprise an award to reimburse the claimant's costs. Sub-Clause 1.9 [*Delayed Drawings or Instructions*] and Sub-Clause 4.7 [*Setting Out*], however, go further than this under the provision that the award should comprise *'Cost plus reasonable profit'*.
11. It is therefore the Contractor's assertion that due to the circumstances entitling him to an extension of the time for completion of the Works, he is also entitled, pursuant to both the Contract and common law, to an award of money to recompense him for the additional costs incurred, plus reasonable profit, as a result of the additional time he has been obliged to remain on site.

Chapter 8

12. The remainder of this section therefore demonstrates the calculation of the additional payment to which the Contractor considers himself entitled.

The Basis of the Calculations for Additional Payment

The Period of Prolongation

1. The Contract Commencement and Completion Dates were as follows:
   a. Commencement date: 3 January 2011
   b. Completion date: 31 December 2012
2. The completion date was subsequently revised by the Engineer through two Extension of Time Awards as follows:
   a. Extension of Time Award No. 1 for Claim No. 1 for Unforeseeable Physical Conditions, which resulted in a substantial increase in the piling and foundation works, revised the completion date to 21 January 2013
   b. Extension of Time Award No. 2 for Claim No. 4, for the late nomination of the facade contractor, revised the completion date to 9 February 2013
3. Thus, the completion date at the time of the delay event that is the subject of this claim was that established by Extension of Time Award No. 2, i.e. 9 February 2013.
4. Section 6 herein has established the Contractor's entitlement to a revised completion date of 11 March 2013.
5. The period of prolongation for which the Contractor has incurred additional costs is therefore:

a. Extension of Time No. 2 Completion Date:	9 February 2013
b. Revised Completion Date for this claim:	11 March 2013
c. Prolongation period:	30 Calendar Days

Costs

1. As demonstrated in the above section, entitled *The Contractor's Entitlement to Additional Payment*, the Contractor is entitled to additional payment equating to *Cost plus reasonable profit* for the additional time that he was obliged to maintain his resources due to the extension of time. Under the principles of recovery, the aggrieved party is entitled to be put back in the position in which it would have been had the contract been performed as originally envisaged. The additional payment for the Contractor's prolongation costs should therefore be based on the costs incurred at the time the damage occurred or, in other words, during the prolongation period. As demonstrated in Section 6 herein,

the time when the damage occurred was between mid January and mid February 2013.

2. The Contractor's monthly report for January 2013[43] includes schedules of staff, plant, transport, equipment and other resources that were deployed at the time in question, and these schedules have been used as a basis for the additional-payment calculations.

3. As demonstrated in the above section, entitled *The Contractor's Entitlement to Additional Payment*, the Contract provides that the additional payment due to the Contractor should include *'reasonable profit'*. The Contractor submits that reasonable profit should be based upon profit achieved previously within the company. Appendix H contains the audited account summary sheets for the past three years, which show that the following percentages of profit have been achieved:

Financial year ending 4 April 2010:	3.24%
Financial year ending 2 April 2011:	4.18%
Financial year ending 1 April 2012:	3.76%
TOTAL:	11.18%
Divide by years:	3 years
Average percentage per year:	3.73%

4. The Contractor submits that this is a fair and reasonable means of establishing the percentage of reasonable profit to be added to the net costs of prolongation and has included this in the calculations contained herein.

5. Appendix I herein contains detailed calculations of the costs incurred as further described below.

Site Establishment Costs

1. Due to the extended time for completion of the Works, the Contractor has been obliged to maintain his resources on site for a period in excess of the time contemplated in the Contract. The Contractor is therefore entitled to be compensated for the additional costs incurred in doing so. Such site resources include site staff, site establishment, transport, plant and equipment.

2. The Contractor's site staff includes site management, site supervision, site administration and non-productive personnel employed on a time-related basis, such as drivers, storekeepers and security personnel.

---

[43] Exhibit 17 – Schedules from JCG's Monthly Report, dated 31/01/13

3. The costs included in employing staff include salaries, insurance, holiday pay, pension contributions and all other costs incurred in employing such personnel. The actual staff costs are substantiated and cross-referenced to payroll data which is included within Appendix J herein.

4. The Contractor's site establishment includes toilet and washing facilities, the Contractor's and Engineer's site offices, office equipment, IT equipment, office furniture, office consumables, workshops, stores, water, electricity, telephones, internet facilities, survey equipment, waste disposal and maintenance of the foregoing.

5. In the case of the Contractor's own site-establishment equipment, this may be used on more than one project during its useful life. Hence, the claimed costs have been based on the monthly depreciation value of each item over the period of prolongation. This represents the estimated monthly cost of the item to the Contractor until the owned piece of equipment is either sold or is no longer usable.

6. Transport costs include for company vehicles provided for the Contractor's staff, site-dedicated vehicles for the movement and collection of materials and the costs of fuel for such vehicles. The costs incurred for such items are either hire costs or depreciation costs as described above.

7. Plant and equipment employed on a time-related basis during the prolongation period include generators, scaffolding, access equipment, barriers, safety and security equipment, small tools and equipment and the cost of fuel and/or maintenance as appropriate.

8. The costs incurred for plant and equipment are either hire costs or depreciation costs. Equipment owned by the Contractor is not charged at purchase cost because it may be used on more than one project during its useful life. Hence, the costs claimed are based on the monthly depreciation value of each item over the period of prolongation. This represents the estimated monthly cost of the item to the Contractor until the owned piece of equipment is either sold or is no longer usable.

9. Appendix I-1 herein contains detailed calculations of the Contractor's site-establishment costs for the prolongation period in the sum of £148,942. The calculations include cross-references to invoices, payroll data and the like which have been included in Appendix J as substantiation of the rates and prices used in the calculations.

Contractual Costs

1. Due to the extended time for completion of the Works, the Contractor has incurred additional costs towards maintaining certain contractual requirements, including insurance for the Works and equipment, insurance against injury to persons,

insurance against damage to property, and the performance bond. The Contractor is entitled to be compensated for the additional costs expended during the period of prolongation for such items.

2. Appendix I-2 herein contains detailed calculations of the Contractor's contractual costs for the prolongation period in the sum of £3278. The calculations include cross-references to invoices and the like which have been included in Appendix J as substantiation of the costs used in the calculations.

Head-Office Overheads and Profit

1. The income from any project contributes not only to the costs of running the project itself, but also to the costs of running the off-site offices of the Contractor.
2. The Contractor, having resources locked into this project during the prolongation period, has lost the opportunity of using those resources on other sites where they would have earned a contribution to the cost of running the head office.
3. The following formula is known as 'Emden's Formula' and is accepted as a recognised method for calculating such costs.

   The formula is in two stages as follows:

   Stage 1:

   $$\frac{\text{Annual OH overhead cost and profit}}{\text{Annual turnover}} \times 100 = \text{H\%}$$

   Stage 2:

   $$\text{H\%} \times \frac{\text{Contract Sum}}{\text{Contract Period (days)}} \times \text{Period of Delay (days)}$$

   $$= \text{Amount recoverable}$$

4. The formula set out above thus notionally ascribes to the Contract an amount in respect of overheads and profit, proportional to the relationship that the value of the Contract bears to the annual turnover of the organisation. Appendix H herein includes the summaries from the Contractor's audited accounts for the last three years from which the head-office overheads and profit have been ascertained.
5. The calculation of these additional costs is shown in Appendix I-3 in the sum of £149,401.

Finance Costs

1. Due to the fact that completion of the project was delayed, the release of retention was deferred for the period of prolongation. The Contractor was therefore obliged to pay finance costs on the retention amount through this additional period.

Chapter 8

2. Payment certificate No. 22 for work executed during October 2012 was paid in January 2013, which is the time that delay occurred.[44] The payment certificate shows that retention was withheld at the maximum amount of 10% of the contract sum, i.e. £5,018,251.

3. Appendix A contains a letter from the Contractor's bank that confirms the interest rate of 4.25% charged during the period in question.[45]

4. The financing costs are calculated as follows:

Retention withheld:	£ 5,018,251
Interest rate per annum:	4.25%
Interest per annum:	£ 213,276
Divide by days per annum:	365 days
Interest per day:	£ 584
Multiply by prolongation period:	30 days
Finance costs for prolongation period:	£ 17,520

Summary of Prolongation Costs

1. A summary of the above is contained in Appendix I-5 and for ease of reference is reproduced as follows:

Description	£
Site-Establishment Costs	148,942
Contractual Costs	3,278
Head-Office Overheads and Profit	149,401
Finance Costs	17,520
Subtotal:	319,141
ADD: Allowance for reasonable profit @ 3.73%	11,904
Total Additional Payment Due:	331,045

2. The Contractor therefore claims that, pursuant to the Contract and to common law, he is entitled to an additional payment of £331,045 in compensation for costs expended throughout the period of prolongation, including reasonable profit as demonstrated herein.

---

[44] Exhibit 9 – Payment Certificate No. 22, dated 28/11/12
[45] Exhibit 28 – National Bank letter, dated 02/04/13

Checklist – Claim for additional payment

1. Cause and effect for additional payment
2. Entitlement to additional payment
3. Nature of the additional payment claimed
4. Measured works
5. Disruption
6. Site-establishment costs
7. Contractual costs, i.e. insurances, bonds and guarantees
8. Head-office overheads and profit
9. Finance costs
10. Profit on costs
11. Basis of evaluation
12. Method of evaluation
13. Calculations
14. Substantiation of the costs used in the calculations

Checklist – Costs to be considered for prolongation

The following comprises an additional checklist of items that should be considered within a typical prolongation-cost claim.

The Contractor's Site Staff

1. Site management
2. Site supervision
3. Site administration
4. Non-productive personnel employed on a time-related basis such as drivers, storekeepers and security personnel

The Contractor's Site Establishment

1. Off-site staff and workers' accommodation
2. Toilets
3. Site offices
4. Workshops and stores
5. Local-authority charges on temporary buildings
6. Utilities: water, electricity, internet and telephone costs
7. Office furniture and equipment
8. Office consumables and stationery
9. Survey equipment
10. Waste disposal
11. Maintenance of the above
12. Insurance

Transport

1. Vehicles for the Contractor's staff
2. Vehicles for the movement and collection of materials, i.e. pickups, vans etc.

Chapter 8

3. Vehicles for the transportation of workers and staff to and from the site
4. Fuel for vehicles

Plant and Equipment

1. Generators
2. Compactors
3. Rollers
4. Site transport for material handling
5. Compressors
6. Pumps
7. Breakers
8. Loaders
9. Excavators
10. Lifting and hoisting equipment
11. Scaffolding
12. Barriers, safety and security equipment and the like
13. Small tools and equipment
14. Temporary lighting and power
15. Fuel and maintenance

Calculations

In a claim that comprises or includes additional payment, the calculations of the amount claimed are obviously a very important part of the claim document. For this reason, the calculations should not be a haphazard collection of numbers, but a logically presented document that provides the reviewer with full understanding of the logic adopted in the calculations. The same principles that apply to the narrative apply equally to the way the calculations are presented. They need to be user-friendly, the logic should be explained and the calculations should be included in a stand-alone format. Invoices, payroll information and other such data should be presented as substantiation, and the calculations should be cross-referenced to the substantiating documents, in much the same way as exhibits are presented in the narrative.

It is usually appropriate to include the calculations of costs or additional payment on spreadsheets and to include these in the appendices. I have seen many methods of producing such calculations and some of these have been so complicated that it has been almost impossible to follow the calculations without sitting down with a calculator and attempting to ascertain the logic used by the person who compiled them. The fact that they have been offered without any explanatory notes has not helped to lessen the confusion. The same principle of ensuring that documents are presented in a stand-alone and user-friendly manner should apply equally to the calculations. The spreadsheets should be titled and the titles should reflect the same titles used in the narrative. Dates and revision numbers of each spreadsheet should be included in case revisions are subsequently required as a result of negotiations with the other party. Page numbers should be included and items should be enumerated or referenced. Column headings should clearly describe the content of the column.

Spreadsheets are an excellent way of producing quite complicated calculations and this is a definite advantage, but if the logic of such calculations cannot be followed by a reviewer, they can become a distinct disadvantage. For this reason, the calculations should be shown in a step-by-step manner with all the logic made evident. To illustrate this, here are two examples of a calculation of the daily cost of hiring a generator:

Example No. 1

Monthly generator hire cost as invoice No. 1234 from Dallaglio Plant Hire	500.00
Daily cost	16.44

In the above example, the reviewer is left to work out for himself how the daily cost was ascertained.

Example No. 2

Description	Calculation	Unit	Amount
Monthly generator hire cost as invoice No. 1234 from Dallaglio Plant Hire		£	500.00
Months per year:		Months	12
Yearly cost:	£500 × 12 months	£	6000.00
Days per year:		Days	365
Daily cost:	£6000 per month/365 days per year	£	16.44

Chapter 8

While the second example demands a little more work, it clearly demonstrates how the cost has been ascertained and leads the reviewer to a logical conclusion. It is obvious which example the reviewer would prefer to be presented with.

Another common flaw exhibited by those who compile spreadsheets is to fail to show the units that the numbers are supposed to represent. This leaves the reviewer with the frustrating task of attempting to work out whether the numbers are supposed to represent monetary values, days, hours, linear, square or cubic metres, litres, percentages or any combination of the above. A number without a unit represents nothing at all and this makes calculations without units worthless.

While we have the ability to produce complicated spreadsheets to demonstrate all sorts of calculations – and in some cases this may be an appropriate thing to do – the traditional and familiar 'bill of quantities' type of format will work best in most situations. Here is an example of the calculations for the Contractor's costs of site staff that could be included in the appendices of our example claim submission.

APPENDIX I
COST CALCULATIONS

Date: 19 April 2013
Rev: 0

Note: The 'Cost Ref.' column contains references to the supporting information submitted as substantiation of the costs and which is contained in Appendix J.

Item	Description	Cost Ref.	Quantity	Unit	Rate (£)	Total (£)
	SITE-ESTABLISHMENT COSTS					
	Site Staff					
A	Project Manager: R Hill	J.001	30	Days	164.38	4,931.40
B	MEP Manager: J Lewsey	J.002	30	Days	136.98	4,109.40
C	Architectural and Finishes Manager: J Robinson	J.003	30	Days	123.28	3,698.40
D	MEP Engineer: W Greenwood	J.004	30	Days	95.89	2,876.70
E	Architectural and Finishes Supervisor: K Bracken	J.005	30	Days	95.89	2,876.70

Item	Description	Cost Ref.	Quantity	Unit	Rate (£)	Total (£)
F	Planning Engineer: M Tindall	J.006	30	Days	109.58	3,287.40
G	Senior Quantity Surveyor: B Cohen	J.007	30	Days	123.28	3,698.40
H	Quantity Surveyor: P Vickery	J.008	30	Days	76.71	2,301.30
J	Project Administrator: B Kay	J.009	30	Days	54.79	1,643.70
	Subcontractor's Site Staff					
K	MEP Project Manager: T Woodman	J.010	30	Days	142.39	4,271.70
L	MEP Engineer: N Back	J.011	30	Days	83.27	2,498.10
M	Interior fit-out Site Manager: L Moody	J.012	30	Days	103.96	3,118.80
	Site Staff: To Summary of Site-Establishment Costs					39,312.00

Checklist – Appendices; calculations

1. Appendix reference and title
2. Revision number and date
3. Page numbers
4. Item numbers
5. Column headings clearly describe the column contents
6. Explanatory notes
7. Units clearly annotated
8. Cross-references to substantiating documents

Chapter 8

Chapter 9

The Appendices and Editing

Now is the time that we usually sit back and congratulate ourselves because all the research, the collecting of evidence and the hard work of writing the narrative has been completed. Unfortunately, however, we still have quite a lot of work to do to put the claim document in a suitable condition for submission. In Chapter 4, we discussed the subjects of making the document user-friendly, ensuring that it is a stand-alone document with the inclusion of exhibits and additional documents to provide substantiation of statements made in the narrative and the costs used in the calculations. Much of this is achieved by the use of, and the organisation of, the appendices.

An effective way to make the document user-friendly is to compile the submission in two volumes, with the narrative contained in the first volume and the supporting documentation in a separate volume or volumes. This allows the reviewer to refer to the supporting documents while reading the narrative contained in the first volume. As discussed in Chapters 5 to 8, the narrative will contain numerous references to exhibits and other documents offered to support the claim and such documents should be separated into appendices and arranged in a logical manner. Each appendix should have dividers with clearly labelled tabs and, if necessary, the appendices should have sub-dividers to assist in the location of documents. For example, if Appendix A contains exhibits referenced 1 to 20, then each individual exhibit should be located behind a sub-divider with an appropriate label from 1 to 20. The inclusion of *all* the information within the claim document will ensure that, as well as being user-friendly, the claim will be able to be reviewed by someone unfamiliar with the project or the circumstances surrounding the claim subject and they will be able to gain a complete understanding of the whole issue without the need for any external references.

Chapter 9

A typical organisation of the appendices for our example claim would be as follows:

APPENDICES

Appendix A: List of Exhibits
Appendix B: EOT 2 Baseline Programme
Appendix C: Revision-C Drawings of the Transformer Room
Appendix D: Photographic Records Showing the Status of the Transformer Room at the Time when the Alteration Work was Instructed
Appendix E: Extracts from the Contractor's Monthly Report for December 2012 Showing the Contractor's Progress Against the Baseline Programme
Appendix F: Revision-D Drawings of the Transformer Room
Appendix G: Impacted As-Planned Programme
Appendix H: Extracts from the Contractor 's Audited Accounts for 2010, 2011 and 2012
Appendix I: Cost Calculations
Appendix J: Supporting Information for the Cost Calculations

The above are arranged in the approximate order in which they appear in the narrative.

The appendices should be shown on the contents page of the claim narrative and also on the volumes containing the appendices. If the appendices are contained in more than one volume, each volume should have an individual contents page detailing the appendices contained in that volume.

Each appendix should include a flysheet behind the divider separating the appendices; the annotation on the flysheet should match the list of contents. An example of a flysheet from our example claim would be as follows:

Chapter 9

APPENDIX D

**Photographic Records Showing the Status of the Transformer Room
at the Time when the Alteration Work was Instructed**

The exhibits should be listed on a separate contents page and either arranged in chronological order or in the order in which they appear in the narrative. If the latter method is adopted, however, it should be noted that if it is necessary to refer to a particular exhibit more than once within the narrative, the order will soon become illogical.

You will note from our claim example that the narrative contains cross-references to the exhibits by the use of footnotes. While it is tempting to give each exhibit an exhibit number in the footnotes at the time of writing the narrative, e.g. *Exhibit 19 – Daily Site Report, 21/11/12*, I can guarantee that if this method is adopted, the exhibit numbers will have to be revised at a later date. Most likely, this will be necessary because, during editing or following a review by another person, it will be necessary to introduce an addition to the narrative that requires additional substantiation, or an additional exhibit will be required to substantiate something or other. When this happens, and especially if the exhibits are arranged in chronological order, any numbers allocated to the exhibits will need to be revised. For this reason it is better to leave the numbering of the exhibits until such time as all editing and internal reviews have been completed and a complete list of exhibits has been established and listed in the List of Exhibits. I find that the least painful way to approach this task is as follows:

Chapter 9

1. Add the footnotes while writing the narrative, but do not at this stage include an exhibit number, for example, *Exhibit X – letter reference 12345, dated 01/02/11*.
2. Compile the list of exhibits when the narrative is complete and arrange these in chronological order.
3. Complete the editing and review process and, if necessary, revise the list of exhibits.
4. Add exhibit numbers to the list of exhibits.
5. Print a hard copy of the list and go back to the narrative and add the exhibit numbers to the footnotes.

Here is the list of exhibits that would be included in our example claim.

APPENDIX A
List of Exhibits

1. The exhibits included herein are submitted to enable the Engineer to verify statements made in the narrative and are cross-referenced to the narrative by the use of footnotes.
2. The exhibits are listed in chronological order and are contained in this appendix under tabbed dividers showing the exhibit number.

Exhibit No.	Reference	Date	Description
1.	Specification No. 23/34/78		Contract Specification Extract
2.	JLD/let/JCG/L-001	29/11/10	Jason Leonard Developments letter
3.	P-1013/Let-0001	03/12/10	Johnson Construction Group letter
4.	ST/JW/Contr/036	09/03/11	Dawson-Wilkinson Partnership letter
5.	DIT No. 042	05/07/11	Dawson-Wilkinson Partnership Drawing Issue Transmittal
6.	ST/JW/Contr/126	22/09/11	Dawson-Wilkinson Partnership letter
7.	ST/JW/Contr/295	17/01/12	Dawson-Wilkinson Partnership letter
8.	ST/JW/Contr/345	05/03/12	Dawson-Wilkinson Partnership letter

Exhibit No.	Reference	Date	Description
9.	Payment Certificate	28/11/12	Payment Certificate No. 22 for work executed to November 2012
10.	P-1013/Let-1345/NEC	02/01/13	Johnson Construction Group letter
11.	Minutes	11/01/13	Site Progress Meeting No. 51
12.	13/0978NC/3945	14/01/13	National Electricity Company Letter
13.	P-1013/Let-1362	15/01/13	Johnson Construction Group letter
14.	P-1013/Let-1367	22/01/13	Johnson Construction Group letter
15.	Minutes	25/01/13	Site Progress Meeting No. 52
16.	P-1013/Let-1373	31/01/13	Johnson Construction Group letter
17.	Monthly Report	31/01/13	Johnson Construction Group Monthly Report
18.	DIT No. 139	04/02/13	Dawson-Wilkinson Partnership Drawing Issue Transmittal
19.	Site Report	05–13/02/13	Daily Site Reports
20.	Site Report	14/02/13	Daily Site Report
21.	Site Report	15/02/13	Daily Site Report
22.	13/0978NC/3973	16/02/13	National Electricity Company letter
23.	P-1013/Let-1421	17/02/13	Johnson Construction Group letter
24.	Site Report	25/02/13	Daily Site Report
25.	Site Report	26/02/13	Daily Site Report
26.	Site Report	27/02/13	Daily Site Report
27.	Site Report	10/03/13	Daily Site Report
28.	Letter	02/04/13	National Bank letter

Chapter 9

Now, for the final checklist to ensure everything has been included:

Checklist – Appendices; general arrangements

1. Claim separated into two volumes
2. Cover pages for each volume

3. Appendices separated by dividers
4. Individual documents and exhibits separated by sub-dividers
5. Dividers labelled
6. Contents page for each volume
7. Flysheets for each appendix containing the appendix title and located behind the dividers
8. List of exhibits or other such documents included at the front of the appropriate appendix
9. Explanation of the order and arrangement of exhibits and other such documents within the appropriate appendix

Editing and Review

We are now at the stage where the narrative has been completed and the appendices have been compiled. While it may be tempting at this point to breathe a sigh of relief, press 'print' and submit the claim document, we still have two important tasks to complete. Firstly, we need to review and edit the whole submission document and secondly, we need to have the exercise repeated by an in-house reviewer. If the in-house reviewer deputised to carry out the review has no knowledge of the project or the circumstances surrounding the claim, then so much the better because he will be reviewing the document from a totally fresh point of view. Consequently, if something does not make sense to him or requires additional explanation, then this subject should be revisited and revised by the author. The in-house reviewer should put themselves in the place of the person who will eventually have the task of reviewing the document and advise the author on unclear passages, incorrect grammar, unsubstantiated statements and the like. The in-house reviewer should also refer to any programmes, calculations and so on that are referenced in the document to ensure that the narrative has incorporated the correct information, that explanations contained in the narrative are easily followed and that any cross-references to other documents are correct. Calculations should also be mathematically checked at this stage.

It is almost inevitable that when the editing and in-house review have been completed, revisions and changes will have to be made. When making such changes it is important to remember to revise *all* sections affected by the change. For example, if it is necessary to change one of the calculations included in the calculation sheets, it will probably be necessary to make corresponding revisions to the narrative in one or more places and probably also in the Executive Summary.

It is usually the case that when a claim is submitted it will be the subject of discussion and used as a basis for negotiations with the other party. Consequently, it is often necessary to produce a revised version of the claim, either to add extra information, or to change something that has been agreed during the negotiation process. In such situations, care must be taken to ensure that the whole of the revised document remains consistent. For example, if a calculation has to be revised which results in a new amount for the claimed additional payment, then the sections of the narrative that deal with the additional payment will need to be revised, in order to maintain consistency with the newly calculated figure, as will the Executive Summary and possibly other sections.

Here is our final checklist to assist with the editing and review process:

Checklist – Editing

1. Has the document been dated?
2. Has the document revision number been included?
3. Have headers and footers been included in the narrative section?
4. If appropriate, have headers and footers been included in the documents incorporated in the appendices?
5. Have page numbers been included in the narrative?
6. If appropriate, have page numbers been included in the documents incorporated in the appendices?
7. Is the line spacing correct and consistent?
8. Are the margins adequate for notes to be made?
9. Are the fonts consistent?
10. Are the alignment and justification of the text consistent?
11. Is the format of section heading, subsection heading and the like consistent?
12. Is the numbering of sections consistent?
13. Are the titles of the contents page, section headings and appendices consistent?
14. Have the exhibits referred to in the narrative been cross-referenced to the appendices?
15. Is the narrative presented in an easily understood and free-flowing writing style?
16. If abbreviations or acronyms are used, have they been defined or explained?
17. Is it clear which party is being referred to at all times within the narrative?
18. Are quotations properly and consistently identified?
19. Are quotations an exact replication of the original?

Chapter 9

20. Are references to other sections of the narrative and the appendices correct?
21. Are the titles of documents included in the appendices consistent with the narrative?
22. Are dates and timelines quoted in the narrative consistent with the programmes included in the appendices?
23. Are calculations that are included in the narrative mathematically correct?
24. Are figures that are quoted in the narrative consistent with the calculations in the appendices?
25. Is the Executive Summary consistent with the rest of the narrative?

Chapter 10

Claim Responses and Determinations

So far, this book has been concerned with presenting a claim on the basis that the claim document should convince the other party, or at least the party responsible for reviewing the claim that, on the balance of probabilities, the claimant has entitlement under the contract and/or at law. There can only be three outcomes of any claim: it will be accepted in full (although this is rarely the case); it will be accepted but with a reduced amount; or it will be rejected entirely. Regardless of the outcome, it will always be someone's job to produce a response to the claim and to prepare a document that justifies it. Sub-Clause 20.1 [*Contractor's Claims*] of FIDIC provides that the following procedure should be adopted on the receipt of a claim by the Engineer:

Within 42 days after receiving a claim or any further particulars supporting a previous claim, or within such other period as may be proposed by the Engineer and approved by the Contractor, the Engineer shall respond with approval, or with disapproval and detailed comments. He may also request any necessary further particulars, but shall nevertheless give his response on the principles of the claim within such time.

And:

The Engineer shall proceed in accordance with Sub-Clause 3.5 [*Determinations*] to agree or determine (i) the extension (if any) of the Time for Completion (before or after its expiry) in accordance with Sub-Clause 8.4 [*Extension of Time for Completion*], and/or (ii) the additional payment (if any) to which the Contractor is entitled under the Contract.

Construction Claims & Responses: Effective Writing & Presentation, Second Edition. Andy Hewitt.
© 2016 John Wiley & Sons, Ltd. Published 2016 by John Wiley & Sons, Ltd.

Sub-Clause 3.5 [*Determinations*] goes on to say:

Whenever these Conditions provide that the Engineer shall proceed in accordance with this Sub-Clause 3.5 to agree or determine any matter, the Engineer shall consult with each Party in an endeavour to reach agreement. If agreement is not achieved, the Engineer shall make a fair determination in accordance with the Contract, taking due regard of all relevant circumstances.

The Engineer shall give notice to both Parties of each agreement or determination, with supporting particulars. Each Party shall give effect to each agreement or determination unless and until revised under Clause 20 [*Claims, Disputes and Arbitration*].

Thus, the Engineer is obliged to approve the claim, or alternatively, to disapprove it with detailed comments. If he finds that he does not have sufficient information to reach a satisfactory conclusion, he must comment on the principles of the claim and request further particulars to enable him to do so. The Engineer is also obliged to attempt to reach agreement with both parties and, failing this, he must make a fair determination on the matter, and his determination should include supporting particulars. The Engineer, or his equivalent under other forms of contract, can thus be regarded as being between a rock and a hard place. On the one hand, if he awards too little, he risks the matter being elevated to a dispute by the Contractor and his findings examined by a dispute board or arbitrators and, on the other hand, if he awards too much, the Employer (who incidentally pays his fees) will regard him as having not protected the Employer's interests, which usually means paying as little for the project as possible. It can be seen from this that a determination has to be equally as well justified as the claim itself and, therefore, the same principles apply equally to response and determination writing as they do to the preparation of claims.

While the above provisions from FIDIC set out a clear procedure which obliges the Engineer to act fairly in producing a determination, it is true to say that not all responses to claims will be treated in this way, and a strategy to deal with the claim should be considered carefully by giving due consideration to the following:

1. Is it the responsibility of the reviewing party to defend the respondent's interests and minimise the claim as much as possible, or to produce a fair and reasonable determination?
2. If the defence of the respondent's interests is of primary concern, does the value of the claim justify the expenditure of a significant amount of

resources to offer up a rigorous defence? Similarly, if a claim is likely to result in a significant award, it is probably worth providing the necessary resources to ensure that a high-quality effort is made.

3. What are the strengths and merits of the claim and its chances of success? Are the odds of success favourable enough to justify the effort and expense in providing a rigorous defence?

4. What is the quality of the submission? Does it provide enough information to enable a proper determination to be made? Does it satisfy the basic principles necessary to establish entitlement and quantum? Can it be properly understood? If the claim does not fulfil the necessary criteria, should it be rejected on that basis, or should the reviewer proceed anyway and attempt to close the matter?

5. The strategy should also consider how the response is to be pitched. Is it felt that the best result would be obtained by minimising any awards and leaving plenty of room for negotiation, or would it be better to ensure that the value of the award is reasonable, that all arguments are absolutely sound and the case for the respondent is as bulletproof as possible? The latter usually results in an initially higher award to the claimant, but is often harder to refute. It is also true to say that while a response strategy aimed at minimising the quantum might have a chance of success if reviewed by inexperienced parties, if it subsequently proceeds to a dispute, such a response is unlikely to succeed when the experts get involved.

6. If a strategy of rigorous defence is adopted, what is the dispute procedure and what is likely to be the outcome if the matter does progress to a dispute?

7. Would the respondent's interests be best served by protracting the matter or by dealing with it in a timely manner? If the former strategy is adopted, would protraction cause the respondent to be in breach of any contractual obligations and, if so, to what effect?

8. Some claims are complicated in their very nature and, if this is the case, they require a certain amount of specialist knowledge and experience to prepare an adequate response. Do the respondent's resources contain the required experience and knowledge to produce the desired result, or should additional resources be brought in?

9. Past and future relationships between the parties should be considered, possibly at executive level, before embarking on a course of action that could end in contention.

10. The actual personnel who are likely to receive the response or determination should also be considered. Will they be difficult to persuade? Has animosity crept into the relationship? Is the person likely to have sufficient knowledge to understand the matter in question and the contractual principles relied upon? Is the claimant likely to engage the services of an expert to assist him, or to deal with the matter on his behalf?

Chapter 10

It must be borne in mind at this point that the standard of proof applicable to civil proceedings is that the case must be judged on the balance of probabilities and that, therefore, if a fair and reasonable determination is to be produced, it would be unreasonable to seek a higher standard of proof than that which would be required by an arbitrator or the courts.

As we have discussed in the claims section of this book, a professionally presented, well-structured submission that covers the essential elements of cause, effect, entitlement and substantiation stands a better chance of success than a poorly presented one. Additionally, from the point of view of a reviewer, such a claim is much easier to deal with and can, consequently, usually be processed to the point of a response (whether positive or negative in conclusion) within a reasonable time frame. On the other hand, a badly presented claim poses several problems for the reviewer who should think carefully before moving on to the next step. The reviewer has basically three options in such a situation:

1. Reject the claim on the grounds that the claimant has not proved his case. Obviously the respondent has to provide and substantiate reasons for such a rejection in his response.
2. Respond with a request for the claimant to provide additional particulars in order that a proper review may be undertaken. In such a case, it would be necessary to advise the claimant exactly what additional particulars are necessary and/or which aspects of the claim do not provide sufficient information for a determination to be made.
3. Produce a determination that is based not only upon the claim submission, but also on the respondent's own knowledge of the matter and additional records available to the respondent.

On several occasions, on receipt of a very poor claim submission, it has been my responsibility, not only to produce a fair and reasonable determination, but also to ensure that the matter is closed in a timely manner. Consequently, due to the inability of the claimant to produce a reasonable claim submission, I have been obliged to adopt the last option. In such cases, and bearing in mind that my determination would be subject to scrutiny by both the Contractor and the Employer, I have found myself having to start from scratch and ensure that my determination considered all the aspects of the claim that the Contractor failed to deal with. Obviously, this can be a lot of work but, depending on the circumstances, it might be worth the effort if it produces a determination that might be agreed upon by the parties and settled in a timely manner. I have to say, though, that this approach tends to fly in the face of the principle that the burden of proof is on the claimant and I cannot envisage a situation where the lawyer for the defence takes the time to

Chapter 10

help the prosecution prepare their case because, as presented, it might not convince the jury. In many cases, however, our job in the construction industry is to ensure that claims are settled in a proactive manner and, in the circumstances, this could well be the correct course of action to take. A distinct disadvantage to this approach is that, if the reviewer adopts this course of action, it is very likely that the Contractor will continue to submit poor-quality claims with the expectation that the reviewer will continue to do his job for him.

Over the years, I have developed a system of reviewing and responding to claims that works well in most situations and complies with common contractual requirements to produce a fair assessment and to consult with the parties. The procedure is as follows:

1. Carry out an initial review of the claim and advise the claimant of any shortcomings that prevent a conclusion from being reached or any additional particulars that are required. This provides the claimant the opportunity to revise the claim to take into account the comments or to provide the required information.
2. After the receipt of additional particulars, or preferably a revised submission, carry out an assessment based upon all the information presented. If at this stage the conclusion is that an award may not be made, or a lesser amount is due because of shortcomings in the claim, then at least the claimant has been provided with a reasonable opportunity to rectify the situation before a conclusion has been reached. The assessment should be presented in a response document in accordance with the principles discussed in this chapter.
3. Issue the assessment to the Employer and the Contractor for review and comment. The important thing here is that the Employer is brought into the procedure at this stage and has an opportunity to provide input.
4. After a suitable time for the parties to review the assessment, convene a meeting with the Employer and the Contractor to discuss the assessment and to accept comments or additional information. Depending on the feedback during the meeting, it may be necessary to request either party to follow up with a formal response and possibly additional information.
5. If necessary, revise the assessment to take into account the comments from the parties. It may be necessary at this point to address comments from the parties in detail and this information should be included in the revision.
6. Issue the revised assessment to the parties for review and comment again and if necessary repeat items 4 and 5 again until there are no additional comments, or at least until such a stage that all meaningful comments have been made.

Chapter 10

7. Once the parties have been given adequate opportunity to provide input, issue the final assessment. If the parties cannot agree at this stage, it may be necessary to issue a formal determination of the matter.

While to a large extent the claim response strategy will dictate the *nature* of the response or determination – that is, whether it is to be rigorously defended or to be determined impartially – this should not affect the principles of how the response document is prepared and presented. As with claim submissions, a well-presented, well-structured and user-friendly determination document will go a long way to persuading the claimant that the reviewer is experienced in such matters, has confidence in his findings and that, consequently, the response has merit and is based on correct principles. At the risk of repeating what has already been discussed at length in the chapters on claim presentation, the following principles should be borne in mind when compiling a response. These principles are especially important if there is a possibility that a determination will be opened up by an external party or at a higher level, possibly as a result of the matter being raised as a dispute.

The response or determination document should demonstrate the professionalism of the company presenting it and should, consequently, include good-quality files or folders; covers and spine labels showing the necessary information; clearly tabbed and labelled dividers; good-quality paper; user-friendly and attractive page layouts with good graphics; headers and footers showing the name of the party who 'owns' the document; the title of the document; page numbers; the document date and, if necessary, the revision number. Documents included as exhibits should be presented in a logical order, be clearly and easily identifiable and cross-referenced. The narrative and supporting appendices and exhibits should, if possible, be contained in separate volumes and sections that are logical and clearly labelled.

The writing style of the narrative should be grammatically correct and correctly punctuated; it should also flow, be easily readable and be properly understood. Simple and direct language is important in this respect in order to provide a proper understanding of the points being made. Abbreviations and acronyms should be avoided unless they are in common usage within the industry and would be understood by anyone not familiar with the project. At the very least, any acronyms that are used should be defined the first time that they are used in the document or in a table of definitions. Care must be taken to avoid ambiguities by the use of words such as 'them', 'they', 'him' and 'it' when referring to the parties, organisations or people. It is preferable to use the actual names or the contractual designations of the parties such as 'the Employer' or 'the Contractor', to avoid any misunderstanding.

It is often a good idea to reproduce whole clauses from the Contract or other documents within the narrative, in order that the reviewer is accurately informed of the provisions. Additionally, the use of certain wording from the Contract may be used within the narrative to good effect when explaining a particular condition (examples of this are given in Chapter 3). In such cases, the use of quotations should be identified by the appropriate use of quotation marks.

As with claim documents, it is important to make the document user-friendly by the use of a suitably large font with line spacing at 1.5 or 2 and large margins to enable a reviewer to make notes in the document as he reads the narrative. Statements made or counter-arguments within the review must also be substantiated by reference to the project records or the like. Such records should be properly referenced within the narrative; copies should be included in the submission document as exhibits; and such exhibits should be clearly arranged and labelled for easy reference. Bear in mind that it is entirely possible that someone who is not intimately familiar with the project in general or the circum-stances of the claim may be called upon to review the document and that, consequently, it is important that the response should be a stand-alone document that contains all the information for such a person to understand the response without reference to any source other than the claim itself. The principle that a document presented in such a way intimates that the party that has produced it has confidence in the response and is ready to submit the document to the next level holds true for responses and determinations as well as for claims.

The format of a response or determination document will, to a large extent, be dictated by the format and quality of the claim document itself. In many cases, it is easier for a reviewer to prepare a response to a comprehensive, well-ordered and logical claim, because the respond-ent can simply follow the same order in which the claim is presented and either concur with the points made or, by including appropriate explanations, establish points of disagreement and the reasons thereof. If a soft copy of the claim can be obtained, it is sometimes a good strat-egy to reproduce parts of the narrative of the claim within the response and add the reviewer's responses, comments and arguments, section by section, in the appropriate places. If this method is adopted, and provided that it is made very clear, perhaps by the use of different fonts or the like, which narrative belongs to which party, then it is possible to create a very user-friendly response or determination document.

On the other hand, if the claim is of poor quality, does not deal with all the necessary issues or is disorganised in presentation, it is usually better to create a totally independent, well-organised and logical response that deals with all the issues raised by the claimant plus anything pertinent to the matter that may have been omitted by the

Chapter 10

claimant. As with a claim document, and especially if the claimant has not included such issues, the response should explain the background of the project, the parties, the form of contract and other issues relevant to the matter. If the claim is subsequently elevated to a dispute, it is possible that the parties responsible for determining the dispute, however hard they try to remain impartial, will, if only subconsciously, have more sympathy for the party that makes their task easier and exhibits professionalism in their presentations, than for the party who makes life difficult by presenting them with a poor-quality document that is difficult to follow or to understand.

It is often necessary to include supporting documents, such as programmes and calculations, within a response or determination, in order to demonstrate and justify the reviewer's opinion or determination on certain aspects of the claim. As with claim presentations, it is important to ensure that such documents, especially if they have been prepared by others, are coordinated with the narrative and to offer explanations as to how they have been prepared and the logic behind them, in such a way that these supporting documents may be easily understood by a reviewer.

Finally, it is always good practice to have someone, preferably a person with no knowledge of the matter, review the response or determination document to ensure that all matters may be properly understood by someone in a similar position.

As discussed previously, the object of a response or determination document is to record the findings following an examination of a claim, which could range from acceptance of the claim in its entirety to outright rejection, but usually falls somewhere between these two extremes. Whatever the result, the findings, especially in the case of a determination, need to be demonstrated, substantiated and justified so as to bring the matter to a close without progressing to another round of responses and possibly to a dispute. As with claims, in order to achieve such a situation, it is important to ensure that the essential elements of cause, effect, entitlement and substantiation (CEES) are adequately examined and dealt with in the response, even if they have not been included in the claim itself.

In Chapter 5, we created an example scenario to demonstrate how the principles of cause, effect, entitlement and substantiation could be used in a typical compilation of a claim for an extension of time. We will now look at how a determination of this claim could be written in accordance with the same principles. The scenario used for this example is repeated here as follows:

1. The Employer has entered into a contract with a construction company to construct 85 two-storey, detached dwellings. The contract is the FIDIC *Contract for Building and Engineering Works designed by the Employer 1999 Edition*.

2. The Employer has also engaged a separate contractor to construct the infrastructure for the development, which includes mains drainage, electricity distribution, telephone ducting, roads, pavements and public landscaping.
3. The two contracts are running concurrently and the Engineer is the same party for both contracts.
4. On 1 February 2010, the infrastructure contractor excavated a trench for a road crossing across the road leading to four of the dwellings and prevented the Contractor from accessing these dwellings.
5. The infrastructure contractor completed the road crossing and backfilled the trench on 9 February 2010.
6. The Contractor considers that the lack of access caused by the road crossing delayed the work to the four dwellings and that this delay had a direct effect on the completion of the project and, consequently, he is entitled to an extension of time for this delay.

For ease of reference, we will assume that the determination may be written by way of a reproduction of sections from the claim to which the reviewer's comments have been added. The original claim sections are presented in Times New Roman font, thus, and the Engineer's comments and determination are shown in `Courier font, thus`.

Please note that, for brevity, the following example is not a full claim document and only contains the sections that deal with cause, effect and entitlement. Also note that the exhibits used as substantiation are only added where they have been included within the Engineer's response.

The Cause

The Contractor's Claim

1. On 1 February 2010, the Contractor wrote to the Engineer to advise that the Employer's infrastructure contractor had excavated a trench across the access road leading to house numbers 36, 38, 40 and 42. These houses may only be accessed by way of the road that was affected by the infrastructure contractor's works. Photographs taken on the same day are included [`in the Contractor's claim`] under Appendix A and show the extent of the infrastructure contractor's work and the restricted access.
2. It was recorded in the site meeting held on 10 February 2010 that the infrastructure contractor's work was completed and access to the four affected houses was re-established on 9 February 2010.

The Engineer's Response

1. While the Engineer concurs with the above, the Contractor's attention is drawn to the Site Coordination Meeting of 19 January 2010 which was attended by the Contractor. The minutes of the meeting[1] were issued to the Contractor on 21 January 2010[2] and under Item 3.6 state the following:

   '*R and S Contractors* (the infrastructure contractors) *to start road-crossing across Road 7 on 1 Feb. Work anticipated to take 3 days*'

2. The Engineer therefore considers that the Contractor was provided with adequate notice to make arrangements to ensure that his work could continue between 1 and 3 February 2010 when access to the dwellings was anticipated to be restricted.

3. The Engineer further considers that it would also have been both prudent and reasonable for the Contractor to assume that the infrastructure contractor's work could extend past the anticipated 3-day period and make contingency plans for such an event.

The Effect

The Contractor's Claim

1. Appendix B [of the Contractor's claim] contains a site plan which has been marked up to show the location of the road crossing. The plan also shows that alternative access to the dwellings in question was not possible due to the location of existing boundary walls and other completed construction works. The photographs contained in Appendix A [of the Contractor's claim] also demonstrate that the construction works shown on the site plan had already been constructed at the time in question. Thus, the road crossing restricted the Contractor's access to the four dwellings by preventing vehicles from reaching the houses to deliver the construction materials necessary for progress to be maintained.

---

[1] Appendix 1 – Coordination Meeting No. 17 minutes, 19/01/10
[2] Appendix 2 – Transmittal No. 1276, 21/01/10

The Engineer's Response

1. While the Engineer concurs with the Contractor's comments made with regard to the access to the dwellings, the Engineer also considers that given the advance notice provided to the Contractor, arrangements should have been made by the Contractor to deliver sufficient materials to the working area to progress the work during the time that the road-crossing works were planned, plus a reasonable contingency against any delays by the infrastructure contractor.

The Contractor's Claim

1. The infrastructure contractor commenced excavation of the road crossing on 1 February 2010, which effectively restricted access from this day. The daily site report of 1 February 2010 and the photographs included in Appendix A [of the Contractor's claim] record that the progress of the individual dwellings was as follows;

   a. House No. 36 – reinforcement to the raft foundation was in progress and due to be completed on 2 February 2010.
   b. House No. 38 – ready for concrete to be poured to the raft foundation.
   c. House No. 40 – blockwork to ground-floor external walls and partitions was in progress and due to be completed on 3 February 2010.
   d. House No. 42 – blockwork completed to ground floor. First-floor precast concrete flooring beams due for delivery and placement on 1 February 2010.

The Engineer's Response

1. The Engineer concurs with the progress recorded by the Contractor.

The Contractor's Claim

1. Progress to the affected dwellings was dependent on the delivery of ready-mixed concrete, concrete blocks, cement, sand, precast concrete flooring beams and other materials to the working areas. No other access route for such vehicles was available due to other activities and completed construction works in this area of the site. The effect of the excavation of the road crossing by the infrastructure contractor was to prevent deliveries from being made between the dates of 1 and 9 February 2010 and,

consequently, to suspend the construction activities until 10 February 2010, the day after the access was reinstated on 9 February 2010. The Daily Site Report of 9 February 2010 and the photographs included in Appendix C [of the Contractor's claim] record the progress of the affected dwellings at the time when the road crossing was reinstated and the works were able to recommence as normal.

The Engineer's Response

1. As stated previously, the Engineer considers that adequate notice was given to the Contractor that the road crossing would affect the access to the dwellings in question between the dates of 1 and 3 February 2010. Consequently, the Contractor should have made arrangements to ensure that sufficient materials were delivered to the work area to ensure that work could continue for the 3 days in question, plus a reasonable contingency against delays by the infrastructure contractor. The Engineer considers that contingency measures to allow for one additional day in this respect would have been reasonable.

2. The Contractor has made extensive use of concrete pumps on the project and the Engineer therefore considers that ready-mixed concrete could have been delivered to the required workface at House Nos. 36 and 38 by means of a concrete pump situated at the far side of the road crossing to the dwellings. The Engineer would have considered awarding additional payment for such a contingency measure.

3. The Engineer concurs that the road crossing was reinstated on 9 February 2010 and that work was able to be recommenced on the affected dwellings by the Contractor on 10 February 2010.

4. The Engineer accepts that the road crossing should have been reinstated by the infrastructure contractor on 3 February 2010, thus allowing the Contractor full access on 4 February 2010. The additional time taken by the infrastructure contractor that prevented the Contractor from continuing work until 10 February 2010 comprises a delay of 6 days.

5. As stated previously, however, the Engineer also considers that the Contractor should have reasonably planned for possible additional delay by the infrastructure contractor by ensuring that adequate materials were available for one additional day, thus ensuring that it was possible to continue work on 4 February 2010 and reducing the effect of the delay to 5 days.

6. Given the fact that the Contractor should have made arrangements to provide materials so that work could continue between 1 and 3 February 2010, the advised dates of the road closure, plus one additional day for contingencies, the Engineer considers that only the effect of the infrastructure contractor's delay in reinstating the road crossing from 5 to 9 February 2010 should be considered as a period during which the Contractor was prevented from carrying out his works to the affected dwellings.

The Contractor's Claim

1. This period of suspension had the effect of delaying the Time for Completion of the Works and this has been demonstrated by impacting the event on the individual activities on the current baseline programme (included [in the Contractor's claim] under Appendix A), in order to produce an impacted baseline programme which is included [in the Contractor's claim] under Appendix B.

The Engineer's Response

1. The Engineer agrees to the appropriateness of the Contractor's method of demonstrating the delay, but not on the amount of delay. The Engineer has therefore created a new impacted baseline programme which is included herein under Appendix 3.

The Contractor's Claim

The effect on each dwelling is shown as follows:

House No. 36

Chapter 10

1. Progress on 1 February 2010: reinforcement to the raft foundation in progress and due to be completed on 2 February 2010.
2. The concrete gang can only complete one raft per day and House No. 38 was programmed to start prior to House No. 36. Thus, concreting to House No. 38 took place on 10 February 2010 when the access was reinstated. The gang followed on with House No. 36 on 11 February 2010. Thus, the effect was to delay concreting of the raft foundation from the planned date of 3 February to 11 February 2010, a delay of 8 calendar days.
3. The effect on the overall programme has been demonstrated by impacting the baseline programme as follows:
   a. Activity: concrete to raft foundation.
   b. Activity start date deferred by 8 calendar days.

The Engineer's Response

1. The Engineer considers that reinforcement could have continued and been completed as planned on 2 February 2010.
2. The Engineer considers that the Contractor could have utilised a concrete pump (for which the Engineer would have considered additional payment) in order to concrete the raft on 3 February 2010 as planned.
3. There is a 3-day curing period before block-work may be commenced, so the Engineer accepts that, allowing for the weekend of 6 and 7 February 2010, materials for this activity would not have been required until 8 February 2010. This is after the time that the Contractor could have anticipated that the access would be prevented by the road crossing.
4. The Engineer accepts that it was not reasonable for the Contractor to plan for additional delay past 4 February 2010 and therefore delivery of materials for the activity 'Blockwork to ground floor' was delayed until 10 February 2010. Thus, the start of this activity was delayed from 8 February until 10 February 2010, a delay of 2 calendar days.
5. The effect on the overall programme has been demonstrated by impacting the baseline programme as follows:
   a. Activity: blockwork to ground floor.
   b. Activity start date deferred by 2 calendar days.

The Contractor's Claim

House No. 38

1. Progress on 1 February 2010: ready for concrete to be poured to the raft foundation on that day.
2. The effect was to delay concreting of the raft foundation from the planned date of 1 February to 10 February 2010, a delay of 9 calendar days.
3. The effect on the overall programme has been demonstrated by impacting the baseline programme as follows:
   a. Activity: concrete to raft foundation
   b. Activity start date deferred by 9 calendar days

The Engineer's Response

1. The Engineer considers that the Contractor could have utilised a concrete pump (for which the Engineer would have considered additional payment) in order to concrete the raft on the planned date of 1 February 2010.
2. There is a 3-day curing period before block-work may be commenced so the Engineer accepts that materials for this activity would not have been required until 5 February 2010, which is later than the time that the Contractor could have anticipated that the access would be prevented by the road crossing.
3. The Engineer accepts that delivery of materials for the activity 'Blockwork to ground floor' was outside the Contractor's control until 10 February 2010 and that, consequently, the contractor was delayed from 5 to 10 February 2010, a period of 5 calendar days.
4. The effect on the overall programme has been demonstrated by impacting the baseline programme as follows:
   a. Activity: blockwork to ground floor
   b. Activity start date deferred by 5 calendar days

The Contractor's Claim

House No. 40

1. Progress on 1 February 2010: blockwork to ground-floor external walls and partitions was in progress and due to be completed on 3 February 2010.

2. The effect was to suspend progress on the ground-floor block-work from 1 February to 10 February 2010 and thus prevent completion until 12 February 2010, a delay of 9 calendar days.
3. The effect on the overall programme has been demonstrated by impacting the baseline programme as follows:
   a. Activity: blockwork to ground floor
   b. Activity duration increased by 9 calendar days

The Engineer's Response

1. As discussed previously herein, the Engineer considers that the Contractor, having been given adequate notice of the road crossing, should have ensured that enough materials were delivered to the work area, to ensure that production continued between 1 and 4 February 2010, the planned period of the road crossing plus one day for contingencies. Consequently, the ground-floor blockwork to external walls and partitions could have been completed as planned on 3 February 2010.
2. Additionally, the Engineer's site diary, extracts from which are included under Appendix 2 herein, notes that blockwork to House No. 40 was ongoing on the morning of 1 February 2010 but had ceased on 2 February 2010 due to lack of materials.
3. The follow-on activity for the blockwork is delivery and installation of the first-floor precast concrete flooring beams which were due to be delivered and installed on 4 February 2010. The Engineer considers that it was not possible to deliver and erect the flooring until 10 February 2010, the day after the reinstatement of the road crossing.
4. The effect on the overall programme has been demonstrated by impacting the baseline programme as follows:
   a. Activity: PCC flooring
   b. Activity start date deferred by 6 calendar days

The Contractor's Claim

House No. 42

1. Progress on 1 February 2010: blockwork completed to ground floor. First-floor precast concrete flooring beams due for delivery and placement on 1 February 2010.

2. The effect was to delay completion of the first-floor precast concrete flooring from 1 February to 10 February 2010, a delay of 9
calendar days.
3. The effect on the overall programme has been demonstrated by
impacting the baseline programme as follows:
   a. Activity: PCC flooring
   b. Activity start date deferred by 9 calendar days

The Engineer's Response

1. The Engineer accepts that, due to lack of
   access for delivery vehicles and a mobile
   crane, the Contractor was unable to carry out
   this activity until the day after the road
   crossing was reinstated.
2. The effect was therefore to delay completion
   of the first-floor precast concrete flooring
   from 1 February to 10 February 2010, a delay
   of 9 calendar days.
3. The effect on the overall programme has been
   demonstrated by impacting the baseline programme as follows:
   a. Activity: PCC flooring
   b. Activity start date deferred by 9 calendar
      days

The Contractor's Claim

1. Reference to the baseline programme included [in   the
   Contractor's claim] under Appendix A shows that this
   cluster of four dwellings was the last to be started and thus the
   last to be completed. The effect of this delay event therefore had
   a direct effect on the Time for Completion of the project. The
   impacted baseline programme included [in the Contractor's
   claim] under Appendix B demonstrates that the effect on the
   individual activities of the affected dwellings has had the overall
   effect of delaying the Time for Completion by 9 days, i.e. until 6
   August 2010.

The Engineer's Response

1. The Engineer concurs that the dwellings
   affected had a direct effect on the Time for
   Completion.
2. The Engineer is of the opinion that any float
   built into the baseline programme should be
   used for the benefit of the project and that
   a delay to the Time for Completion should not
   occur until the entire float has been used.

The Engineer notes that the Contractor has retained float attached to certain of the dwellings in his impacted baseline programme and that this has had the effect of extending the Time for Completion past the date that it would have been if the float had been removed.

3. The Engineer's impacted baseline programme, which is included in Appendix 3 herein, has impacted the activities as described in the Engineer's responses above and has removed the float. This programme demonstrates that the effect on the Time for Completion is to extend it by only 2 days to 30 July 2010.

The Contractor's Entitlement

Extension to the Time for Completion

The Contractor's Claim

1. The Contractor's entitlement to an extension to the Time for Completion is contained within the provisions of Sub-Clause 8.4 [*Extension of Time for Completion*] which provides that:

   '*The Contractor shall be entitled subject to Sub-Clause 20.1* [Contractor's Claims] *to an extension of the Time for Completion if and to the extent that completion for the purposes of Sub-Clause 10.1* [Taking Over of the Works and Sections] *is or will be delayed by any of the following causes:*

   …

   (e) any delay, impediment or prevention caused by or attributable to the Employer, the Employer's Personnel, or the Employer's other contractors on the Site.'

2. The event of the road closure by the infrastructure contractor as described herein clearly falls under the provision of '*delay, impediment or prevention caused by or attributable to the … Employer's other contractors on the Site*'. Consequently, the Contract provides that the Contractor shall be entitled to '*an extension of the Time for Completion if and to the extent that completion … is or will be delayed*'. The claim submitted herein contains the Contractor's request for an extension of Time for Completion for the 9 days of delay demonstrated herein.

The Engineer's Response

1. The Engineer concurs that the Contractor is entitled, pursuant to Sub-Clause 8.4 [*Extension*

of Time for Completion], to an extension to the Time for Completion.

2. The Engineer, however, considers that the Contractor should have taken appropriate measures to continue with the work to the affected houses between 1 and 3 of February 2010, plus one additional day that should reasonably and prudently have been allowed as a contingency against delay by the infrastructure contractor.

3. The above opinion is supported by Sub-Clause 8.1 [*Commencement of Works*] which provides that '*The Contractor … shall then proceed with the Works with due expedition and without delay*'.

4. The Engineer also considers that, in a contract such as this, where there is a specific requirement for the work to be completed within a specific time, then time is of the essence. In such a situation, an implied term exists within the contract that the parties must do everything reasonably possible to comply with the time stated.

5. The Engineer therefore considers that the Contractor, having being notified of the road crossing some 13 days before the road crossing was commenced, failed to ensure that sufficient materials were delivered to the working area, failed to make alternative arrangements to deploy a concrete pump to ensure that work could continue, did not act '*with due expedition*', or pay due regard to the implied term within the Contract to do everything reasonable to mitigate delay.

6. The Engineer therefore concludes that the Contractor is entitled to an extension to the Time for Completion only for the effect of the denied access due to the road crossing from 5 to 10 February 2010, which, as demonstrated herein, had the effect of delaying the completion date by only 2 calendar days and not by the 9 calendar days claimed by the Contractor.

Additional Payment

The Contractor's Claim

1. The above sub-clause contains a reference to Sub-Clause 20.1 [*Contractor's Claims*] which provides that:

   '*If the Contractor considers himself to be entitled to any extension of the Time for Completion and/or any additional payment, under any*

Clause of these Conditions or otherwise in connection with the Contract, the Contractor shall give notice to the Engineer, describing the event or circumstance giving rise to the claim. The notice shall be given as soon as practicable, and not later than 28 days after the Contractor became aware, or should have become aware, of the event or circumstance.'

2. The above sub-clause provides that the Contractor may claim additional payment. Due to the circumstances causing a delay to the Time for Completion, the Contractor was obliged to pay for labour and plant that was obliged to stand idle during the time that access to the work area was denied. Secondly, the Contractor was obliged to remain on site for a period greater than was originally intended and thereby incurred additional costs in maintaining his site establishment, providing finance for the works and maintaining head-office overheads. He was also prevented from earning a contribution from other projects through having his resources tied up for the extended period.

3. The principles of recovery where one party to a contract has defaulted are well established in law. Essentially, the aggrieved party is entitled by an award of money to be put back in the position in which it would have been had the contract been performed as originally envisaged.

4. Sub-Clause 20.1 [*Contractor's Claims*] also provides that '*Within 42 days after receiving a claim ... the Engineer shall respond with approval, or with disapproval and detailed comments*' and subsequently that '*Each Payment Certificate shall include such amounts for any claim as have been reasonably substantiated as due under the relevant provision of the Contract*'.

5. It is therefore the Contractor's further claim that due to the circumstances entitling him to an extension of the Time for Completion of the Works, the Contractor is also entitled, pursuant to both the Contract and to the law, to additional payment to recompense him for the costs incurred as a result of labour and plant standing idle and the additional time he has been obliged to remain on site. The Contractor's claim in this respect will be submitted by way of a separate claim.

The Engineer's Response

1. The Engineer concurs that the Contractor is entitled to additional payment to compensate him for the costs incurred as a result of idle labour and plant time, provided that this is properly demonstrated and the effect does not

include the time for each dwelling for which the
Contractor should have mitigated.

2. The Engineer also concurs that the Contractor is
   entitled to additional payment for the additional
   time he has been obliged to remain on site, but considers that the period of prolongation, upon which
   such additional payment should be based, is 2 calendar days and not 9 calendar days as claimed by the
   Contractor.

3. The Engineer awaits the Contractor's separate claim
   in this respect.

Liquidated Damages or Penalties

The Contractor's Claim

1. The Employer's entitlement to deduct penalties for late completion is
   contained under Sub-Clause 8.7 [*Delay Damages*] as follows:

   *'If the Contractor fails to comply with Sub-Clause 8.2 [Time for
   Completion], the Contractor shall subject to Sub-Clause 2.5 [Employer's
   Claims] pay delay damages to the Employer for this default'*

2. This sub-clause refers to Sub-Clause 8.2 [*Time for Completion*] which
   provides that:

   *'The Contractor shall complete the whole of the Works, and each Section
   (if any), within the Time for Completion for the Works or Section (as the
   case may be) including:*

   a. achieving the passing of the Tests on Completion, and
   *b. completing all work which is stated in the Contract as being required
      for the Works or Section to be considered to be completed for the pur-
      poses of taking-over under Sub-Clause 10.1* [Taking Over of the
      Works and Sections].'

3. As has been examined earlier herein, the Time for Completion may,
   however, be extended under the provisions of Sub-Clause 8.4 [*Extension
   of Time for Completion*] as follows:

   *'The Contractor shall be entitled subject to Sub-Clause 20.1 [Contractor's
   Claims] to an extension of the Time for Completion if and to the extent
   that completion for the purposes of Sub-Clause 10.1 [Taking Over of the
   Works and Sections] is or will be delayed.'*

4. Thus, the entitlement of the Employer to the payment of delay damages
   is negated for any circumstances that entitle the Contractor '*to an extension of the Time for Completion*'. As is demonstrated herein, the Contractor
   is entitled to such an extension of time and therefore the Employer is not
   entitled to the payment of delay damages by the Contractor.

Chapter 10

The Engineer's Response

1. The Engineer concurs that the Employer is not entitled to the payment of delay damages for the period
   of the extension of time, but considers that the
   period of the extension of the Time for Completion
   is 2 calendar days and not 9 calendar days as
   claimed by the Contractor.

<u>*Conditions Precedent to Entitlement*</u>

The Contractor's Claim

1. Sub-Clause 20.1 [*Contractor's Claims*] provides that:

   '*If the Contractor considers himself to be entitled to any extension
   of the Time for Completion and/or any additional payment, under
   any Clause of these Conditions or otherwise in connection with the
   Contract, the Contractor shall give notice to the Engineer, describing
   the event or circumstance giving rise to the claim. The notice shall be
   given as soon as practicable, and not later than 28 days after the
   Contractor became aware, or should have become aware, of the event
   or circumstance.*

   *If the Contractor fails to give notice of a claim within such period of
   28 days, the Time for Completion shall not be extended, the Contractor
   shall not be entitled to additional payment, and the Employer shall be
   discharged from all liability in connection with the claim. Otherwise, the
   following provisions of this Sub-Clause shall apply.*

   *The Contractor shall also submit any other notices which are required
   by the Contract and supporting particulars for the claim, all as relevant to
   such event or circumstance.*

   *The Contractor shall keep such contemporary records as may be neces-
   sary to substantiate any claim, either on the Site or at another location
   acceptable to the Engineer. Without admitting the Employer's liability, the
   Engineer may, after receiving any notice under this Sub-Clause, monitor the
   record-keeping and/or instruct the Contractor to keep further contemporary
   records. The Contractor shall permit the Engineer to inspect all these
   records, and shall (if instructed) submit copies to the Engineer.*

   *Within 42 days after the Contractor became aware (or should have
   become aware) of the event or circumstance giving rise to the claim, or
   within such other period as may be proposed by the Contractor and approved
   by the Engineer, the Contractor shall send to the Engineer a fully detailed
   claim which includes full supporting particulars of the basis of the claim and
   of the extension of time and/or additional payment claimed.*'

2. The requirement to give notice of entitlement to an extension of time
   and to describe the event giving rise to the claim within 28 days of
   the event is a condition precedent to the Contractor's entitlement. The

Contractor is also obliged to submit a fully detailed claim of the extension of time claimed within 42 days of the event.

3. The Contractor submitted a notice of claim on 15 February 2010, which is within the 28-day period prescribed in the Contract. The submission contained herein comprises the detailed claim and supporting particulars, thereby satisfying the provisions of this sub-clause.

4. The Contractor has therefore complied with the conditions of this sub-clause and is consequently entitled to an extension of the Time for Completion until 6 August 2010.

The Engineer's Response

```
1. The Engineer concurs that the Contractor has com-
   plied with the conditions precedent contained in
   the Contract and that the Contractor is therefore
   entitled to an extension of the Time for Completion,
   but for the reasons stated herein, this should
   only be extended until 30 July 2010.
```

<u>Conclusion</u>

The Contractor's Claim

1. The following is a summary of the Contractor's entitlement as discussed in this section:

2. The Contractor is entitled under the Contract to an extension to the Time for Completion for delay, impediment or prevention caused by or attributable to the Employer's other contractors on the Site.

3. Due to the circumstances entitling the Contractor to an extension of the Time for Completion, the Contractor is also entitled, pursuant to both the Contract and common law, to additional payment for the costs incurred as a result of the additional time he has been obliged to remain on site. The Contractor's claim in this respect will be submitted by way of a separate claim.

4. The Employer is not entitled to the payment of delay damages by the Contractor.

5. The Contractor has complied with the conditions precedent to entitlement.

The Engineer's Response

```
1. The Engineer concurs that:
   a. The Contractor is entitled to an extension to
      the Time for Completion.
   b. The Contractor is entitled to additional pay-
      ment due to the additional time for which the
      Contractor has been obliged to remain on site.
```

Chapter 10

```
        c. The Employer is not entitled to the payment
           of delay damages by the Contractor.
        d. The Contractor has complied with the con-
           ditions precedent to entitlement.
     2. The Engineer, therefore, for the reasons stated
        herein, considers and determines pursuant to
        Clause 3.5 [Determinations] that the Contractor
        is entitled to an extension of the Time for Com-
        pletion of 2 calendar days until 30 July 2010.
```

In the above determination, the method adopted by the Engineer for the response has been facilitated because the original claim was presented in a clear and logical order. It was therefore appropriate to compose the determination by means of the addition of comments and counter-opinions to each section of the claim. This would not be appropriate, however, if the claim was not logically presented or did not include the essential elements of cause, effect, entitlement and substantiation. In such a case, it would be more fitting to create a completely new response document compiled in a logical manner, similar to the example used in Chapters 6–8. This would enable the essential elements to be addressed properly. In such a case, it would be necessary to make frequent references to the claim document either by page and paragraph numbers, or preferably by the reproduction of sections from the claim within the response narrative. Whichever method is adopted, the important thing to remember is that the response absolutely must deal with the essential elements of cause, effect, entitlement and substantiation, even if the claim itself does not do so.

Finally, here is a suggested layout and order of a response document. This may also be used as a checklist to ensure that everything has been considered and included, either in the claim itself, or in the response, whichever is appropriate:

Front Cover

1. Claimant's name
2. Project title
3. Claim title or brief description
4. Revision reference
5. Revision date
6. Company logo
7. The names of the parties
8. Claim number

9. Volume number
10. Document reference number
11. Author
12. Reviewer

Contents

1. Section numbers
2. Section titles
3. Page numbers

Executive Summary

1. Contains a summary of all sections
2. Includes any last-minute changes and revisions to the main narrative

Background to the Response or Determination

1. Brief details of the Contract
2. Details of the parties
3. Brief details and description of the Project
4. Brief details of the submitted claim
5. Details of the Contract procedure for claim submissions and determination
6. Final or interim claim

Definitions, Abbreviations and Clarifications

1. Definitions and abbreviations of the parties
2. Contractual definitions
3. Method of dealing with quotations
4. Method of dealing with cross-references
5. Arrangement of the response document

The Contract Particulars

1. Details of the parties – the Employer
2. Details of the parties – the Contractor
3. Details of the parties – the Engineer
4. Details of the parties – other relevant parties
5. The form of contract
6. The applicable law
7. The Tender Date
8. The Contract Sum
9. The Commencement Date

10. The Completion Date
11. Previous extension-of-time awards
12. Milestone dates
13. The relevant conditions of contract

The Method of Delay Analysis

1. The suitability of the method of delay analysis used
2. The programme and method used by the claimant to demonstrate the delay and the effect on the Time for Completion and its correctness
3. The Baseline Programme

Claim for an Extension to the Time for Completion

1. The Cause
2. The Effect
3. Delay Analysis
4. Examination of how the claimant's delay analysis has been created and its suitability and accuracy
5. Revised Time for Completion
6. Concurrent delays
7. Entitlement to an extension of time under the Contract
8. Examination of conditions precedent
9. Examination of entitlement in cases where claimant has not complied with conditions precedent
10. Examination of the Employer's entitlement to delay damages
11. Substantiation by reference to the project records

Claim for Additional Payment

1. Cause and effect for additional payment
2. Entitlement to additional payment
3. Nature of the additional payment claimed
4. Measured works
5. Disruption
6. Site-establishment costs
7. Contractual costs, i.e. insurances, bonds and guarantees
8. Head-office overheads and profit
9. Finance costs
10. Profit on costs
11. Basis of evaluation
12. Method of evaluation
13. Calculations
14. Substantiation of the costs

Appendices

1. Exhibits
2. Delay analysis
3. Calculations

Appendices; Calculations

1. Appendix reference and title included
2. Revision number and date included
3. Page numbers included
4. Item numbers included
5. Column headings clearly describe the column contents
6. Explanatory notes included
7. Units clearly annotated
8. Cross-references to substantiating documents included

Chapter 10

While this book is primarily concerned with the preparation and review of claims, it is also appropriate to consider what happens in situations where the parties cannot agree on the matter, and the claim is elevated into a dispute.

Most contracts have provisions whereby the Engineer, or his equivalent under other forms of contract, is required to make a fair determination of the claim; they also include a requirement that the parties attempt to reach amicable agreement in situations in which either party does not accept the Engineer's determination. If such agreement is not reached, then the contract usually provides a further procedure whereby the issue is referred to mediation, conciliation, arbitration or other form of dispute resolution and ultimately, of course, the parties may find themselves in litigation. Sub-Clause 3.5 [*Determinations*] of FIDIC has this to say on the subject of attempted agreement and the Engineer's determination:

Whenever these Conditions provide that the Engineer shall proceed in accordance with this Sub-Clause 3.5 to agree or determine any matter, the Engineer shall consult with each Party in an endeavour to reach agreement. If agreement is not achieved, the Engineer shall make a fair determination in accordance with the Contract, taking due regard of all relevant circumstances.

The Engineer shall give notice to both Parties of each agreement or determination, with supporting particulars. Each Party shall give effect to each agreement or determination unless and until revised under Clause 20 [*Claims, Disputes and Arbitration*].

Construction Claims & Responses: Effective Writing & Presentation, Second Edition. Andy Hewitt.
© 2016 John Wiley & Sons, Ltd. Published 2016 by John Wiley & Sons, Ltd.

The traditional way of settling disputes is to refer the matter to arbitration with the rules of arbitration being described in the contract. A typical arbitration procedure requires each of the parties to propose one suitably qualified arbitrator and for the two appointees to appoint a third member who is often delegated the responsibilities of the chairman of the arbitration panel. Given that the three arbitrators should be experts, to some extent at least, on the subject of the dispute and the construction industry in general, one would expect that most issues should be able to be reviewed, and an arbitrator's determination issued fairly quickly. This is not usually the case, however, because it is necessary at this stage to agree on the members of the arbitration panel who then need to be appointed formally by the parties. Rules and procedures have to be established; the arbitrators need to have time available to devote to the dispute; and they have to take the time to become familiar with the parties, the project and the issues surrounding the dispute. Given all this, it is easy to see why it may take several months before the arbitrators are able to even consider the matter of the actual dispute. More often than not, lawyers for both parties become involved somewhere during this process and, consequently, we now have a new set of people with a different level of understanding of the construction process adding their opinions to the mix. One must also consider that it may be to the advantage of one of the parties to delay the outcome of the arbitration and, in such a case, deliberate tactics may be adopted to delay and obfuscate the whole process. Of course, if, following the arbitration, the matter progresses to court proceedings, then lawyers definitely will need to become involved and the various proceedings, possibly including the additional involvement of expert witnesses, will serve to extend the time for the resolution of the dispute, quite possibly to years rather than months. It has been said that arbitration was a perfectly good process until it was hijacked by lawyers. While this may be a somewhat cynical point of view, it is certain that if matters do progress to this level, the resolution of the dispute will become both very protracted and extremely costly.

The industry recognised that such a situation was not desirable and that it provided little benefit to the project itself and, consequently, what was needed was a relatively easy procedure to settle disputes in a timely and cost-effective manner, in order that the parties could henceforth devote their energies to completing the project. One solution proposed was the appointment of dispute boards, otherwise known as dispute adjudication boards, dispute review boards or combined dispute boards, all of which are essentially the same thing. FIDIC, under Sub-Clause 20.2 [*Appointment of the Dispute Adjudication Board*], has this to say on dispute boards:

Disputes shall be adjudicated by a DAB in accordance with Sub-Clause 20.4 [*Obtaining Dispute Adjudication Board's Decision*]. The Parties shall jointly appoint a DAB by the date stated in the Appendix to Tender.

The DAB shall comprise, as stated in the Appendix to Tender, either one or three suitably qualified persons ("the members"). If the number is not so stated and the Parties do not agree otherwise, the DAB shall comprise three persons.

If the DAB is to comprise three persons, each Party shall nominate one member for the approval of the other Party. The Parties shall consult both these members and shall agree upon the third member, who shall be appointed to act as chairman.

However, if a list of potential members is included in the Contract, the members shall be selected from those on the list, other than anyone who is unable or unwilling to accept appointment to the DAB.

The agreement between the Parties and either the sole member ("adjudicator") or each of the three members shall incorporate by reference the General Conditions of Dispute Adjudication Agreement contained in the Appendix to these General Conditions, with such amendments as are agreed between them.

The terms of the remuneration of either the sole member or each of the three members, including the remuneration of any expert whom DAB consults, shall be mutually agreed upon by the Parties when agreeing the terms of appointment. Each Party shall be responsible for paying one-half of this remuneration.

If at any time the Parties so agree, they may jointly refer a matter to the DAB for it to give its opinion. Neither Party shall consult the DAB on any matter without the agreement of the other Party.

Thus, under FIDIC, the parties jointly appoint the dispute board, which may comprise either one or three members. The dispute board may, if necessary, consult with external experts in order to carry out their duties. In addition to referring disputes to the dispute boards for adjudication, the parties may also jointly refer a matter to the board for its opinion.

The dispute board is paid a monthly retainer fee for becoming familiar with the project and the particulars of the contract, and for keeping abreast of progress and events; it is further compensated for actual time spent during site visits and for dealing with matters that are referred to the board. The costs of the dispute board and any experts with whom the board engages are borne equally by both parties.

Sub-Clause 20.4 [*Obtaining Dispute Adjudication Board's Decision*] goes on to say:

If a dispute (of any kind whatsoever) arises between the Parties in connection with, or arising out of, the Contract or the execution of the Works, including any dispute as to any certificate, determination, instruction, opinion or valuation of the Engineer, either Party may refer the dispute in writing to the DAB for its decision, with copies to the other Party and the Engineer. Such reference shall state that it is given under this Sub-Clause.

For a DAB of three persons, the DAB shall be deemed to have received such reference on the date when it is received by the chairman of the DAB.

Both Parties shall promptly make available to the DAB all such additional information, further access to the Site, and appropriate facilities, as the DAB may require for the purposes of making a decision on such dispute. The DAB shall be deemed to be not acting as arbitrator(s).

Within 84 days after receiving such reference, or within such other period as may be proposed by the DAB and approved by both Parties, the DAB shall give its decision, which shall be reasoned and shall state that it is given under this Sub-Clause. The decision shall be binding on both Parties, who shall promptly give effect to it unless and until it shall be revised in an amicable settlement or an arbitral award as described below. Unless the Contract has already been abandoned, repudiated or terminated, the Contractor shall continue to proceed with the Works in accordance with the Contract.

If either Party is dissatisfied with the DAB's decision, then either Party may, within 28 days after receiving the decision, give notice to the other Party of its dissatisfaction. If the DAB fails to give its decision within the period of 84 days (or as otherwise approved) after receiving such reference, then either Party may, within 28 days after this period has expired, give notice to the other Party of its dissatisfaction.

In either event, this notice of dissatisfaction shall state that it is given under this Sub-Clause, and shall set out the matter in dispute and the reason(s) for dissatisfaction. Except as stated in Sub-Clause 20.7 [*Failure to Comply with Dispute Adjudication Board's Decision*] and Sub-Clause 20.8 [*Expiry of Dispute Adjudication Board's Appointment*], neither Party shall be entitled to commence arbitration of a dispute unless a notice of dissatisfaction has been given in accordance with this Sub-Clause.

If the DAB has given its decision as to a matter in dispute to both Parties, and no notice of dissatisfaction has been given by either Party within 28 days after it received the DAB's decision, then the decision shall become final and binding upon both Parties.

FIDIC thus provides that, should either party disagree with the dispute board's decision, then they may proceed to refer the matter to arbitration.

FIDIC also contains general conditions of the dispute-adjudication agreement and procedural rules from which the following provisions are summarised:

1. The agreement is a tripartite agreement between the Employer, the Contractor and the sole member or each of the members comprising the board.
2. The dispute adjudication board may be appointed at the commencement date of the main contract or as soon as practicable after this date.
3. The board is obliged to be impartial.
4. The board is obliged to have relevant experience in the type of work of the project and the interpretation of contract documentation.
5. The board is obliged to disclose any potential conflicts of interest or previous dealings with the parties.
6. The board shall visit the site at regular intervals and is obliged to make itself available for site visits, and to become acquainted with any potential problems or claims, be available for hearings and the like and to give advice and opinions on any matter relevant to the contract.
7. The parties are obliged to provide the board with copies of the Contract, progress reports, and other documents pertinent to the performance of the Contract, and the board is obliged to become conversant with the Contract and the progress of the works.

The above provisions may at first seem very similar to those of arbitration, but there are some very important differences between the two procedures. The dispute board is appointed early in the contract period and is obliged to become familiar and remain acquainted with the project and the Contract by regular site visits and through the review of documentation. By this very process, it is natural that the board will also become familiar with the parties and the project personnel. The fact that the board has already been appointed and is acquainted with the Contract, the site and any areas of potential dispute, ensures that it may deal with disputes and provide adjudication in a much reduced period than would be the case if the matter were referred to arbitration. The dispute board is, in fact, obliged to provide a reasoned decision on a matter that is referred to the board within 84 days. I would suggest that such a period could easily be equal to, or less than, the time it would take to make the necessary arrangements to commence arbitration proceedings, and much less than the time required for the arbitrators to provide a decision.

The Dispute Resolution Board Foundation has gathered statistics over many years from projects on which dispute boards have been employed and these make impressive reading.

Chapter 11

On construction projects having dispute boards, the average number of disputes referred to the boards is 1.2 per project. This is fewer than the average number of disputes taken to arbitration or to court on projects without dispute boards, and this supports the opinion that the very presence of a board that is able to provide relatively quick decisions prevents the submission of spurious claims and unreasonable determinations and reduces posturing by both parties. On projects having dispute boards, an impressive 98% of disputes are resolved at dispute-board level. Additionally, of the 2% of disputes where one of the parties did not accept the board's decision and proceeded to arbitration, almost all of the arbitrations supported the dispute board's decisions. FIDIC, for example, allows a party to proceed to arbitration if it does not agree with the board's decision but, given these statistics, such a party would have to give very serious consideration as to whether to accept the board's decision on the matter or to proceed to arbitration in the hope that a new panel of experts would arrive at a different decision. All this, of course, has an added advantage that the parties can concentrate on construction issues rather than on claims and disputes, with the obvious benefit that this brings to the project.

According to the Dispute Resolution Board Foundation, the cost of employing dispute boards is between 0.05% of the construction cost on dispute-free projects and 0.25% for more difficult projects and it must be remembered that these costs are shared equally between the parties. If these figures are compared with the cost of arbitration, which Dr. Nael Bunni has quoted as being between US$150,000 and US$200,000 per day or, as assessed by FIDIC, may be as much as 10–15% of the project value, I would suggest that if it were possible to insure against disputes at such costs, the parties would be rushing to their insurance brokers in order to do so. Additionally, if the Contractor has comfort in the knowledge that his claims will be dealt with fairly and reasonably and that any disputes will be decided in an impartial and timely manner, it would be reasonable to assume that he would consider that such a situation would remove a certain amount of risk for which he would otherwise need to include within his price. From this, it is also reasonable to assume that projects with dispute-board provisions would result in potentially lower bids.

Another significant advantage available to the project through the use of dispute boards is that in addition to referring disputes to the board, the parties can ask the board to provide an opinion on a matter. If one considers that disputes are often caused by differences in the interpretation of the Contract, an opinion on the correct interpretation by the board could very well head off a potential claim or, alternatively, provide enough confidence in the potential merits of a claim situation to gainfully pursue the matter. As it was once put to the delegates at a

Dispute Resolution Board Foundation seminar which I attended: if three grey-haired father figures, who are experts in their individual fields of construction, advise the parties that, in their opinion and having carefully considered the matter in question, the outcome of an arbitration would be this or that, most people would accept it as being good advice. I can certainly think of several instances in my career, before I became a grey-haired father figure myself, where I would have welcomed such advice.

The advantages of dispute boards have been recognised by many institutions internationally and their usage is becoming more widespread through the world, with significant usage in the USA, where, probably unsurprisingly, given the USA's penchant for litigation, dispute boards have been seen by government agencies as a significant dispute-avoidance tool. The World Bank and other banks such as the European Investment Bank, the European Bank for Reconstruction and Development and the Asian Development Bank now insist on the inclusion of dispute boards in contracts for any project funded by them, as does the European Union. The International Chamber of Commerce also recommends their use. As we have already discussed, FIDIC has incorporated provisions for dispute boards in the 1999 editions of their standard forms of construction contracts, as has the Institution of Civil Engineers.

Given the impressive track record of this form of dispute resolution along with the many advantages associated with dispute boards and the endorsement of so many respected international agencies, it seems that dispute boards are set to make significant contributions to the industry.

Chapter 11

Information Sources

Additional information and reference material may be obtained from the following organisations:

Claims Class
Education and training on construction claims and related subjects
Email: hello@constructionclaimsclass.com
www: constructionclaimsclass.com

The Institute of Construction Claims Practitioners
Email: hello@instituteccp.com
www: http://instituteccp.com

Fédération Internationale des Ingénieurs-Conseils or International Federation of Consulting Engineers (FIDIC)
Email: fidic.pub@fidic.org
www: http://.www.fidic.org

The Dispute Board Federation
www: http://www.dbfederation.org

The Dispute Resolution Board Foundation
www: http://www.drb.org

Dispute Boards MENA
Email: hello@disputeboardsmena.com
www: disputeboardsmena.com

The Society of Construction Law
www: http://www.scl.org.uk

Construction Claims & Responses: Effective Writing & Presentation, Second Edition. Andy Hewitt.
© 2016 John Wiley & Sons, Ltd. Published 2016 by John Wiley & Sons, Ltd.

FIDIC Clause References

Construction Claims & Responses: Effective Writing & Presentation, Second Edition. Andy Hewitt.
© 2016 John Wiley & Sons, Ltd. Published 2016 by John Wiley & Sons, Ltd.

Index

Printed and bound by CPI Group (UK) Ltd, Croydon, CR0 4YY

27/10/2024

14581107-0001